QUANT PHYSICS FOR BEGINNERS

Discover the history to the quantum
physics, learn the principles and
theories in a simple way for
beginners

James B. Martin

Quantum Physics for beginners

Copyright © James B. Martin 2021,

Quantum Physics for beginners

Table of Content

INTRODUCTION

What is Quantum Physics?

Hi, free students! Regardless of whether you chose up this book from plain interest or you're considering taking a class, I trust I can cause you to feel somewhat surer with regards to the ideas of quantum physics. Quantum physics is the study of matter and energy at the most crucial level. It intends to reveal the properties and practices of the very structure squares of nature. While numerous quantum tests inspect tiny items, for example, electrons and photons, quantum peculiarities are surrounding us, following up on each scale. Nonetheless, we will most likely be unable to identify them effectively in bigger items. This might give some unacceptable impression that quantum peculiarities are unusual or extraordinary. Indeed, quantum science closes holes in our insight into physical science to provide us with a more complete image of our regular daily existences. Quantum revelations have been consolidated into our essential comprehension of materials, science, science, and cosmology. These revelations are an important asset for development, leading to gadgets like lasers and semiconductors, and empowering genuine advancement on innovations once viewed as absolutely speculative, like quantum PCs.

Quantum Physics for beginners

Physicists are investigating the capability of quantum science to change our perspective on gravity and its association with reality. Quantum science might even uncover how everything in the universe (or in numerous universes) is associated with all the other things through higher aspects that our faculties can't fathom.

The field of quantum physical science emerged in the last part of the 1800s and mid 1900s from a progression of exploratory perceptions of particles that didn't appear to be legit with regards to traditional physical science. Among the essential revelations was the acknowledgment that matter and energy can be considered as discrete bundles, or quanta, that have a base worth related with them. For instance, light of a decent recurrence will convey energy in quanta called "photons." Every photon at this recurrence will have a similar measure of energy, and this energy can't be separated into more modest units. Indeed, "quantum" has Latin roots and signifies "how much." Information on quantum standards changed our conceptualization of the particle, which comprises of a core encompassed by electrons. Early models portrayed electrons as particles that circled the core, similar as the manner in which satellites circle Earth. Current quantum material science rather comprehends electrons as being disseminated inside orbitals, numerical portrayals that address the likelihood of the electrons' presence in more than one area inside a given reach at some random time. Electrons can hop starting with one orbital then onto the next as they acquire or lose energy, however they can't be found between.

You, me and the gatepost – at some level at any rate, we are in general getting into the quantum rhythm. To clarify how electrons travel through a CPU, how photons of light get gone to electrical flow in a sunlight powered charger or enhance themselves in a laser, or even exactly how the sun continues to consume, you'll need to utilize quantum physical science. If somebody somehow managed to request us the day to day routine or genuine models from Quantum Material science, the vast majority of us would be absolutely ignorant about it. When you get to be aware of the genuine uses of Quantum Material science, you may ponder that the instances of the equivalent were directly before you. Let's take a look at a number of things that depend on quantum physics for their operations.

1.   Fluorescent light : The light which you are getting from the cylinders or those wavy bulbs is a consequence of a quantum peculiarity in particular. In fluorescent lighting, a modest quantity of mercury fume is energized into the plasma. Mercury can radiate light in the apparent reach. Along these lines, the following time you switch on the lights of your room around evening time, ensure you express gratitude toward Quantum Physics.

2.   Computer and mobile phone: The entire Computer world depends on the rule of Quantum Physical science. Quantum physics discusses the wave idea of electrons, and, thus, this structures the premise of the band construction of strong items on which semiconductor-based hardware are fabricated. Not to fail to remember that we can control the electrical properties of silicon simply because we can concentrate on the wave idea of electrons. When the band structure is changed, the conductivity modifies also. How could the band structure be changed? Obviously, Quantum Physical science knows the response!

3.   Transistor: Transistor have inescapable uses and are utilized to intensify or switch

electrical signs and electrical power. Intently checking out the construction of transistors, we would understand that it comprises of layers of silicon related with different components. Computer chips are made by millions these, and these chips structure the force to be reckoned with of the multitude of mechanical contraptions which have become fundamental to human life. Had Quantum Material science not become possibly the most important factor, these chips would not have been made and neither would work areas, tablets, workstations, cell phones, and different devices have tracked down their direction into human existence

4.   Biological compass: Assuming you imagine that majorly mankind has been

fortunate enough to utilize Quantum physical science, you are thoroughly off-base! As per speculations by researchers, birds like European Robin utilize Quantum Material science to relocate. A light-touchy protein called cryptochrome contains electrons. Photons, subsequent to entering the eyes of the bird, hit cryptochrome, and extremists are delivered. These extremists empower the bird to "see" an attractive guide. Another hypothesis proposes that the snouts of the birds contain attractive minerals. Scavangers, reptiles, creepy crawlies, and surprisingly a few warm blooded animals utilize such kind of attractive compass. You may be shocked to know the sort of cryptochrome which is utilized for route by flies has

13

additionally been found in the natural eye! Nonetheless, its utilization is muddled.

5.   Toaster: The bread toast which you appreciate while tasting on your morning tea can advance toward your plate simply because of Quantum Physical science. The warming component of the toaster oven shines red to toast a cut of bread. Toaster ovens are for the most part alluded to as the justification for why Quantum Physical science appeared. The pole in the toaster oven gets hot, which, thus, is liable for toasting the bread.

I really want to enlighten you regarding a more key idea known as the Copenhagen translation. All in all, this guideline existed as a method for overcoming any issues between quantum characteristics and

Classical physics. It's essentially an assortment of perspectives concerning the importance of quantum mechanics and the way that most quantum esteems are viewed as level headed or up to translation by each particular physicist. To the extent is known, the Copenhagen understanding is essentially credited to Niels Bohr and Werner Heisenberg, two early trailblazers of quantum mechanics. Niels Bohr was a Danish physicist most remarkably associated with his revelations in the fields of nuclear construction and quantum hypothesis. He is, obviously, the organizer of the Bohr model of the molecule. This model says that electrons' energy levels scarcely register on the scientific scale; the actual electrons can jump from one 'orbit' of an atomic nucleus to another. He is additionally liable for the idea of complementarity; essentially, this is the possibility that all items have properties that are reciprocal to each other yet aren't really noticeable. One such illustration of this is the connection among position and energy.

One of the most fundamental "facts" of quantum mechanics, as per Bohr, is that one should gauge all qualities of a worth, like the position and energy of an electron, to characterize the article in one's investigations completely. Along these lines, in like manner, you should have every one of the potential qualities of an idea to comprehend it completely. Then again, Werner Karl Heisenberg was a German hypothetical physicist and probably the best brain concerning quantum mechanics. Heisenberg was an understudy of Bohr's, gathering his idea of the vulnerability rule from a large number of Bohr's talks and thoughts. So we know since particles have their wavelike properties—and waves have molecule like properties. Shouldn't something be said about electrons? Amusingly enough, an electron is its own thing—it's undefinable as either a wave or a molecule. This is on the grounds that it is through the estimation of a molecule, wave, or particle, and so forth, that its definition can be found, and in fact, you can't quantify an electron

17

the same way you would some other substance of its sort.

This carries us to the idea of vulnerability and how Werner Heisenberg took the possibility that waves don't exist in space and shaped it to give verification behind the way that on a nuclear level, it's much more troublesome and in some cases difficult to gauge position and energy. You can nail down a molecule's position, however at that point it's incredibly impossible you'll have the option to get its energy—and visa versa. Dissimilar to traditional material science, quantum physical science doesn't manage rigorously substantial responses. As a particular outcome, there are a great deal of questions that should stay a secret.

That is the reason this discipline bargains in probabilities rather than surenesses. To a limited extent because of Bohr's tutelage, Heisenberg was commended as the maker of quantum mechanics, and it was for this creation he was compensated the Nobel Prize in 1932.

Together, Bohr and Heisenberg gathered implications and cycles of quantum mechanics. Most strikingly, Heisenberg squeezed an unmistakable split between the spectator and the framework or article being noticed. Simultaneously, Bohr accepted that the cycles of an article or framework would in any case occur, independent of whether or not they are noticed or meddled with. This prompted the possibility that quantum mechanics itself is on a very basic

19

level immense, that any perceptions or estimations made in one second will be wrong in the following. This could be one reason why such countless individuals these days see material science, quantum or in any case, as alarmingly difficult to reach and befuddling. Assuming that you're one of those individuals, don't be dispirited; there are as yet numerous ideas and practices inside quantum physical science which can make sure about explicit statistical data points. On a verifiable side note, the meaning of Bohr and Heisenberg's gathering is worth focusing on, past what they were at last ready to make together. The two men (Bohr being more established by sixteen years) met in Nazi-involved region. However nor were military men, they knew about the goings-on of the rest of the world. All things considered, they were all the more promptly worried about the development of information and disclosure than the continuous obliteration of human existence. This is honestly stunning, taking into account that Heisenberg was German and a Lutheran, while Bohr was the senior of the two—was known to be a Jew and of Danish extraction. Besides, while in 1932 the two men were formally Nobel laureates, Bohr hid by not really trying to hide as a Jewish man in the yetabandoned Denmark, offering a place of refuge to numerous German Jewish researchers of the day— including one Albert Einstein. However he was apparently never an individual from the Nazi party, Heisenberg wouldn't leave his country and announced himself a Patriot.

Nonetheless, even now in the Nazi development and the destruction of 'non-Aryan' actual ideas or disclosures, both Bohr and Heisenberg (alongside Einstein) advocated the progressive ideas of relativity and vulnerability by then, at that point, these thoughts were viewed as polluted. Besides, Nazi Germans were instructed to adjust to the standards of communist Leninist authoritative opinion set as predominant by Hitler. As ahead of schedule as the 1920s, communism Leninism ruled as the 'official' philosophy of the Socialist Coalition of the Soviet Association and the USSR. In the times of Hitler's administration, it was

22

considered as the portrayal of nature, society, and logical information, just as a method for these ideas to change and develop—as indicated by communist/Leninist socialist qualities. While communism might lecture a particular sort of fairness, in which all property is possessed by individuals and everybody is assigned their portion, be it money related or scholarly, Leninism is somewhat more limit. It sounds much more like Hitler's style. There's a foundation of an autocracy over the regular workers, driven by a more intense, overwhelming party, which in the long run drives society into socialism. Non-Aryan physical science, for example, that spearheaded by Einstein and Bohr—and unexpectedly, Karl

Marx—were said to soil the unrivaled logical intellectualism of the Aryan race. Freud was viewed as coarse, and Einstein named a Jew as well as a 'pacifistic internationalist' (two convictionswhich conflicted with Nazi belief system), had raised relativity; an idea the 'top' patriot researchers couldn't comprehend and along these lines marked as misrepresentation and garbage.

All of this is significant, I guarantee. Both Bohr and Heisenberg were instrumental in advancing the review and comprehension of quantum physical science and mechanics, and it is crucial for note the snags which were crossed over to present to us the logical information we have today. The Copenhagen understanding permitted perceptions to be made of explicit quantum 'frameworks,' just as goals to be made through substantial numerical qualities. This peculiarity, as well, has a name: the wave capacity or condition. The wave condition essentially permits the estimation of likelihood amidst logical perception. At the point when the wave work implodes or changes, its characteristic mathematic esteem (likewise called an eigenstate) becomes clear. On the off chance that all of that befuddled you, consider it thusly: the breakdown of a wave work is in a real sense simply the hypothetical interaction that happens when a quantum actual idea is noticed and noted. Simply the most common way of being seen and investigated changes the worth of the wave work. OK? Numerous years later Bohr and Heisenberg, American physicist Hugh Everett III set off to demonstrate that the wave work is basically the meaning of any molecule's 'real essence'— i.e., their characteristic quantifiable 'esteem.' Notwithstanding, he didn't have confidence in the breakdown of a wave work, in which one end was made which innately changed the noticed article.

Rather than the breakdown of a wave work, he said rather that perception of the thing birthed endless and changed equal universes. By its actual nature, estimation of something gave presence to quite a few real factors in which the article isn't changed by perception—aside from the one Universe, where the wave work has seemed to have imploded. Set all the more forth plainly: whatever can happen will occur. We are simply ready to see each reality or result in turn. Whether or not you think you have any past information on physical science overall or not, you may have some thought of what all that is known as 'traditional' physical science involves. That is all that stuff about gravity, and inactivity, push versus pull, and what occurred at the time Isaac Newton watched an apple tumble from a tree. You may have even concentrated on a tad of traditional materialscience in center everyday schedule school.

However, you may not understand what the critical distinction is between the two. It's in reality beautiful straightforward. Classical Physics concentrates on perceptible cycles, similar to an apple tumbling from a tree, a tire moving down a slope, or those apparent to the unaided eye. It's the investigation of movement, things like speed, removal, and speed increase. It's frequently applied to the investigation of the development of the planets, just as the day by day events in the ordinary existence of movement and power. Quantum physical, on the other hand, deals exclusively with microscopic bodies of matter. Henceforth the prior discuss electrons! Photons (light energy) are a major piece of such a physical science, similar to the speculations of likelihood and—who could have imagined—relativity. Also, it manages the elusive, every one of the particles and cells and portions of cells that the natural eye can't see without assistance and generally don't keep a decent arrangement of guidelines, as the laws of gravity and movement will more often than not do.

Chapter One

Big names and significant dates

Since we have discussed about some of the pioneer scientists of quantum physics have been discussed in the previous chapter, other notable scientists that has contributed to the knowledge of quantum physics will be further discussed in this chapter.

Quantum Physics for beginners

Sir Isaac Newton(1642 – 1727) was an English Physicists.New Scientist once depicted Isaac Newton as "the incomparable virtuoso and most mysterious person throughout the entire existence of science." His three biggest disclosures — the hypothesis of all inclusive attraction, the idea of white light and analytics — are the justifications for why he is viewed as a particularly significant figure throughout the entire existence of science. Newton's hypothesis of widespread attraction says that each molecule in the universe draws in each and every molecule through the power of gravity. The hypothesis assists us with anticipating how protests as extensive as planets and as little as individual impacting atoms will interface; it shows us the manner in which tremors echo through the Earth's covering and how to construct assembling that can endure them. His straightforward condition for all inclusive attractive energy, written in 1666 when he was 23, helped oust in excess of 1,000 years of Aristotelian thinking (supported by Greek space expert Claudius Ptolemy) which said that objects possibly moved assuming an outer power drove that movement. Newton's three laws of movement, distributed 20 years after the fact in his Principia, set up that each article in a condition of uniform movement will stay in that condition of movement except if an outer power follows up on it, that power approaches mass occasions speed increase and that for each activity there is an equivalent and inverse response. These laws were among quick to clarify essential parts of nature with basic numerical recipes that were helpful in a huge scope of genuine situations. Albeit the laws were subsequently supplanted by Albert Einstein's more exact hypotheses about spacetime and general relativity, they laid the preparation for this and any remaining current contemplated material science and the idea of the real world. Newton was likewise quick to comprehend the rainbow, and to refract white light with a crystal into its part tones and back again into white light, setting up unbending test evidence notwithstanding serious analysis from his counterparts. One of the results of his analyses with light was the Newtonian telescope, still broadly utilized today. Newtonian telescopes utilize a reflecting mirror to keep away from the shading bending and rainbow impact besetting telescopes that utilization focal points. At long last, Newton found and characterized analytics, the numerical framework for getting change, which he applied to general physical science.

Depending on the success of Isaac Newton was a man named Humphrey Davy (1778-1829). He was a Cornish chemist popularly known for his contribution to the discoveries of chlorine and iodine.However, as a little fellow, Davy apprenticed for a specialist. Through this work and his own sharp means, he was capable really to show himself science and certain ways of thinking and, indeed, seven dialects. His emphasis was on the substance sythesis of gases, like nitrous oxide and its applications in the clinical field. He, specifically, was one of the main scholastics to support the utilization of nitrous oxide (presently known as "laughing gas") as a sedative during the medical procedure, to lessen the patients' aggravation levels. In his initially distributed work in 1799, Davy disproved another researcher, Antoine-Laurent Lavoisier; not at all like Lavoisier's distributions, Davy contended that the course of hotness is, truth be told, a consequence of movement, and in like manner that the peculiarity of light is a kind of issue. On the off chance that you have any simple information on key material science,

34

you may be familiar with erosion and what happens when you rub two sticks or two bits of sandpaper together: you get heat! Yet, as Davy came to, you may likewise realize that light is made of minuscule particles called photons, which move around in a concentrated electromagnetic field. However not rigorously identified with material science, one more fantastic achievement by Humphrey Davy incorporates that of the disclosure of a few components, for example, what you may see on the periodical table. By adding electrical charges to bits of potash (what you may call Himalayan salt) and sodium carbonate (a substance compound used to relax hard water and make glass), he disengaged potassium and sodium

35

artificially. Through additional experimentation on profoundly antacid soil tests, he likewise created confines of magnesium, calcium, strontium, and barium. Strontium is a great deal like calcium in that it can reinforce human bone design and is one of the basic parts for building new bone. Barium is totally a made component involved today in paints and the enormous boring machines used to observe oil holds inside the Earth, which we then, at that point, use to control our motors and vehicles.

James Clerk Maxwell(1831 – 1879) was a Scottish Physicist and researcher answerable for the old style hypothesis of electromagnetic radiation, which was the principal hypothesis to depict power, attraction and light as various indications of a similar peculiarity. Maxwell's conditions for electromagnetism have been known as the "second extraordinary unification in Physics where the first had been acknowledged by Isaac Newton. James Clerk Maxwell is one of the goliaths of physics. Sadly, his work is less popular than that of different greats – potentially on the grounds that his delegated magnificence – Maxwell's Equations – are so difficult to comprehend. In creating these conditions, he was the very first researcher to bind together any of nature's major powers. He found that power and attraction are really, at the most unimaginable level, a similar power – the electromagnetic power. In doing as such, Maxwell demonstrated that light is an electromagnetic wave thus connected power, attraction, and optics.

As though this accomplishment were sufficiently not, his active hypothesis of gases precisely clarified the beginning of temperature. He brought likelihood into the physical science of the tiny, establishing the framework for quantum hypothesis. He was the very first individual to create a shading photo; and he utilized science to clarify Saturn's rings north of 100 years before the Voyager shuttle affirmed that he was totally correct. He, indeed, was subsequently refered to by Einstein himself as having made ready for Einstein's hypothesis of relativity, which is around fifty years or so later Maxwell's demise, would build up new methodologies on items' mass and energy and how the planets and the space-stuff around them move around. An antecedent to relativity, Maxwell's revelation of electromagnetic fields would likewise lay the primary flagstones for the advancement of innovations, for example, radio, electrical cables, metal indicators, and in the long run, microwaves and even cellphones. These gadgets utilize electric and attractive fields of power to run—to deliver sound, power, and to permit us to converse with each other from huge spans separated by means of phone.

Erwin Schrödinger, (conceived August 12, 1887, Vienna, Austria—Passed on January 4, 1961, Vienna), Austrian hypothetical physicist who added to the wave hypothesis of physics and to different basics of quantum mechanics. He shared the 1933 Nobel Prize for Material science with English physicist P.A.M. Dirac. In 1925,Erwin Schrödinger presented a condition that contributed enormously to quantum mechanics. In straightforward terms, it depicts what electrons do under practically any conditions. Expecting to be that matter (e.g., electrons) could be viewed as the two particles and waves, in 1926 Erwin Schrödinger formulated a wave equation that precisely determined the energy levels of electrons in molecules.

The Schrodinger equation made the convoluted universe of quantum physical science unsurprising in the domain of electrons. He took the Bohr particle model above and beyond. Schrödinger utilized numerical conditions to portray the probability of tracking down an electron in a specific position . This nuclear model is known as the quantum mechanical model of the atom. Not at all like the Bohr model, the quantum mechanical model doesn't characterize the specific way of an electron, yet rather, predicts the chances of the area of the electron. This model can be depicted as a core encompassed by an electron cloud. Where the cloud is generally thick, the likelihood of observing the electron is most noteworthy, and on the other hand, the electron is less inclined to be in a less thick space of the cloud.His most popular complaint was the 1935 psychological test that later became known as Schrödinger's Cat.A Cat is secured a steel box with a modest quantity of a radioactive substance to such an extent that following one hour there is an equivalent likelihood of one iota either rotting or not rotting. In the event that the iota rots, a gadget crushes a vial of harmful gas, killing the feline. In any case, until the crate is opened and the molecule's wave work implodes, the particle's wave work is in a superposition of two states: rot and non-rot. In this way, the feline is in a superposition of two states: alive and dead. Schrödinger thought this result "very ludicrous," and when and how the destiny of the cat has been a subject of much discussion among physicist. Of the relative multitude of physicists of his age, Schrödinger stands apart due to his exceptional scholarly flexibility.

David Joseph Bohm (20 December 1917 – 27 October 1992) was an American-English Physicist who has been depicted as perhaps the main hypothetical physicist of the twentieth century and who contributed unorthodox idea to quantum hypothesis, neuropsychology and the way of thinking of psyche. Bohm began questioning the Copenhagen interpretation in the late 1940s while writing a book on quantum mechanics. According to the Copenhagen interpretation, a quantum entity such as an electron has no definite existence apart from our observation of it. We cannot say with certainty

43

whether it is either a wave or a particle. The interpretation also rejects the possibility that the seemingly probabilistic behavior of quantum systems stems from underlying, deterministic mechanisms. Bohm found this view unacceptable. "The whole idea of science so far has been to say that underlying the phenomenon is some reality which explains things," he explained. "It was not that Bohr denied reality, but he said quantum mechanics implied there was nothing more that could be said about it." Such a view reduced quantum mechanics to "a system of formulas that we use to make predictions or to control things technologically. I said that's not enough. I don't think I would be very interested in science if that were all there was." In 1952 Bohm proposed that particles are indeed particles--and at all times, not just when they are observed in a certain way. Their behavior is determined by a force that Bohm called the "pilot wave." Any effort to observe a particle alters its behavior by disturbing the pilot wave. Bohm thus gave the uncertainty principle a purely physical rather than metaphysical meaning. Niels Bohr had interpreted the uncertainty principle as meaning "not that there is uncertainty, but that there is an inherent ambiguity" in a quantum system, Bohm explained.

Bohm's interpretation gets rid of one quantum paradox, wave/particle duality, but it preserves and even highlights another, nonlocality, the capacity of one particle to influence another instantaneously across vast distances. Einstein had drawn attention to nonlocality in 1935 in an effort to show that quantum mechanics must be flawed. Together with Boris Podolsky and Nathan Rosen, Einstein proposed a thought experiment involving two particles that spring from a common source and fly in opposite directions.

Hans Jenny (August 1904, – June 1972) was a Swiss Physicist and Natural scientists who instituted the term cymatics to depict acoustic impacts of sound wave peculiarities. Wave peculiarities, you may recollect, is the conduct of waves—normally light or sound waves— when they move. Both of those recently referenced kinds of waves can reflect (ricochet off and back), refract (head through and shift bearing), diffract (twist around an obstruction), or submit impedance (consolidate into one, denser wave). Water-sound symbolism, one more of Jenny's innovations—however some authorize Ernst Chladini, the German physicist of the eighteenth century, similar to its actual trailblazer— is more a peculiarity of excellence than a perceivable result of science. The manner in which it works is that sound waves are some way or another presented to water or pushed through the water, shaping undulating designs that are distinctive relying upon the sound recurrence, tone, and pitch of the sound.

Quantum Physics for beginners

For Jenny's situation, it is a physical, substantial portrayal of the force of waves (for this situation, sound).

Nonetheless, what's considerably more amazing with regards to this thought is that the idea was applied to artistic expression—manufacturers of wood instruments like guitars, violins, and cellos shape and art their gadgets with a particular goal in mind to create the ideal sound. The waves ricochet off the bends and the states of an instrument when it's played, and the thicker the 'string,' the more significant the tone. On the off chance that you notice when somebody draws a bow across the most minimal string on a cello, the actual line vibrates; this is basically one simple way people can notice sound waves—similar as waves in clear water.

Chapter Two
Heisenberg Uncertainty
principle

Quantum Physics for beginners

The Heisenberg Uncertainty principle is an actual law that structures part of quantum mechanics. It says that the more exactly you measure the place of a molecule, the less definitively you can know its movement (force or speed). Furthermore, the more exactly you measure a molecule's movement, the less unequivocally you can know its position. This implies that it is impossible to measure the exact position and momentum of a particular object.This is in opposition to our regular experience of life, where these estimations are free of one another and can be estimated as exactly as we'd like. At the point when it was presented in 1927, it took some time for individuals to acknowledge. Heisenberg's uncertainty principle expresses that there is inborn vulnerability in the demonstration of estimating a variable of a molecule. Normally applied to the position and energy of a molecule, the guideline expresses that the more definitively the position is known the more unsure the force is as well as the other way around. This is in opposition to traditional Newtonian physical science which holds all factors of particles to be quantifiable to a subjective vulnerability given adequate gear. The Heisenberg uncertanity Rule is a crucial hypothesis in quantum mechanics that characterizes why a researcher can't quantify various quantum factors all the while. Until the beginning of quantum mechanics, it was held as a reality that all factors of an item could be known to correct accuracy all the while briefly. Newtonian material science put no restrictions on how better strategies and methods could decrease estimation vulnerability so it was possible that with appropriate consideration and exactness everything data could be characterized. Heisenberg made the strong suggestion that there is a lower breaking point to this accuracy making our insight into a molecule innately questionable.

It may not be straightforward ,this rule, however it's considerably harder to clarify it. One way Heisenberg attempted to disclose it to individuals was to say that the demonstration of noticing something influences the result. This standard depends on the wave-particle duality of matter. In spite of the fact that Heisenberg's Uncertainty principle can be overlooked in the perceptible world (the vulnerabilities in the position and speed of items with moderately enormous masses are irrelevant), it holds critical worth in the quantum world. Since atoms and subatomic particles have tiny masses, any increment in the precision of their positions will be joined by an expansion in the vulnerability related with their speeds.

Envision that you are in a research facility, attempting to notice an electron through a magnifying lens, to quantify its position and speed. The light that you are utilizing in this perception ricochets off the electron and arrives

53

at your eyes. That is the means by which you can see it. In any case, the light influences the electron when it skips off it. Light contains minuscule particles called photons, and those particles have a specific measure of energy. This is an amount that quick and weighty items have bunches of: a football player running has a great deal of force, which is the reason it's difficult to stop them. On account of light, its measure of force relies upon the frequency (shade) of the light waves, which can be controlled in a research facility. On the off chance that the sort of light we use for our perception has photons with a great deal of energy, then, at that point, we can without much of a stretch see where the electron is (its position). It resembles focusing a truly brilliant light into the magnifying instrument. But since they have a ton of force, they'll move it to the electron when they skip off it, making it accelerate. This will make it difficult to tell how quick it is moving (its speed). Our perception has impacted the electron's speed.

However, in the event that the sort of light we use has photons with scarcely any force, we can only with significant effort see where the electron is (its position). It resembles involving a magnifying lens in a faint room. In any case, since the photons have little energy, they won't influence the electron's speed, making it more straightforward to realize how quick it is moving (its speed). The better we know the speed, the harder it is to know the position. Furthermore the better we know the position, the harder it is to know the speed. That is the uncertainty rule. It is difficult to envision not having the option to know precisely where a molecule is at a given second. It appears to be instinctive that assuming a molecule exists in space, then, at that point, we can highlight where it is; notwithstanding, the Heisenberg Uncertainity principle plainly shows in any case. This is a direct result of the wave-like nature of a molecule. A molecule is fanned out over space so that there just is definitely not an exact area that it possesses, yet rather involves a scope of positions. Essentially, the energy can't be exactly known since a molecule comprises of a bundle of waves, every one of which have their own force so that, best case scenario, one might say that a molecule has a scope of energy.

We should consider on the off chance that quantum factors could be estimated precisely. A wave that has a totally quantifiable position is fallen onto a solitary point with an endless frequency and along these lines endless energy as per de Broglie's condition. Likewise, a wave with an entirely quantifiable force has a frequency that sways over all space endlessly and accordingly has an endless position. You could do a similar psychological test with energy and time. To definitively gauge a wave's energy would take a boundless measure of time while estimating a wave's accurate occurrence in space

57

would need to be fallen onto a solitary second which would have endless energy. In the field of quantum mechanics, Heisenberg's uncertainty principle is a fundamental theory that explains why it is impossible to measure more than one quantum variable simultaneously. Another implication of the uncertainty principle is that it is impossible to accurately measure the energy of a system in some finite amount of time.

For what reason is it Difficult to Quantify both Position and momentum At the same time?

To show Heisenberg's uncertainty rule, consider a model where the place of an electron is estimated. To quantify the place of an article, a photon should crash into it and return to the estimating gadget. Since photons hold some limited force, an exchange of momenta will happen when the photon crashes into the electron. This exchange of momenta will make the force of the electron increment. Accordingly, any effort to quantify the place of a molecule will expand the vulnerability in the worth of its force. Applying a similar guide to a plainly visible article (say a ball), it very well may be perceived that Heisenberg's vulnerability guideline insignificantly affects estimations in the naturally visible world. While estimating the place of a b-ball, there will in any case be an exchange of force from the photons to the ball. In any case, the mass of the photon is a lot more to modest than the mass of the ball.

Consequently, any energy bestowed by the photon to the ball can be disregarded.

Heisenberg's uncertainty rule forces a limitation on the exactness of synchronous estimation of position and energy. The more exact our estimation of position is, the less precise will be our energy estimation as well as the other way around. The actual beginning of the Heisenberg vulnerability rule is with the quantum framework. Assurance of position by playing out an estimation on the framework upsets it adequately to make the assurance of q uncertain as well as the other way around. We will find out with regards to the guideline exhaustively beneath.

How can we describe Heisenberg's uncertainty Principle?

Heisenberg's uncertainty principle expresses that for particles showing both molecule and wave nature, it won't be imaginable to precisely decide both the position and speed simultaneously. The guideline is named after German physicist, Werner Heisenberg who proposed the vulnerability standard in the year 1927. This standard was defined when Heisenberg was in attempting to fabricate an instinctive model of quantum material science. He found that there were sure key factors that restricted our activities in knowing specific amounts. This guideline fundamentally features that synchronous estimation of position and the speed or force of tiny matter waves will have a mistake to such an extent that the result of the blunder in estimation of position and energy is equivalent or in excess of an indispensable various of a steady.

Beginning of the uncertainty principle.

One of the significant focuses for the beginning of the uncertainty is exclusively because of the double idea of a wave-particle. Each molecule is said to have a wave nature and the likelihood of observing particles is most extreme where the waveforms are the best. In the event that the molecule has more prominent undulation then the frequency turns out to be more undefined or ambiguous. Nonetheless, we can decide of force of the molecule. From what we have perused up until this point we can say that the particles that have unequivocal positions will have no decent speed. Then again, a molecule with an obvious frequency will show a conclusive or exact speed. With everything taken into account, assuming we get an exact perusing of one amount it will just prompt huge vulnerability in the estimation of the other.

Mathematically,

All the more explicitly, assuming that one knows the exact energy of the molecule, it is difficult to know the exact position, as well as the other way around. This relationship likewise applies to energy and time, in that one can't quantify the exact energy of a framework in a limited measure of time. Vulnerabilities in the results of "form sets" (force/position) and (energy/time) were characterized by Heisenberg as having a base worth relating to Planck's steady separated by 4π. All the more obviously.

- $\Delta p\Delta x \geq h/\pi\pi$

- $\Delta t\Delta E \geq h/4\pi$

- Where Δ alludes to the uncertainty in that factor and h is Planck's consistent.

Beside the numerical definitions, one can sort out this by envisioning that the more cautiously one attempts to gauge position, the more disturbance there is to the framework, bringing about changes in energy. For instance look at the impact that estimating the position has on the force of an electron versus a tennis ball. Suppose to quantify these items, light is needed as photon particles. These photon particles have a quantifiable mass and speed, and come into contact with the electron and tennis ball to accomplish a worth in their position. As two items crash into their separate momenta ($p=m*v$), they grant propositions momenta onto one another. At the point when the photon contacts the electron, a piece of its energy is moved and the electron will currently move comparative with this worth relying upon the proportion of their mass. The bigger tennis ball when estimated will have an exchange of force from the photons also, however the impact will be reduced in light of the fact that its mass is a few significant degrees bigger than the photon. To give a more down to earth depiction, picture a tank and a bike crashing into each other, the tank depicting the tennis ball and the bike that of the photon. The sheer mass of the tank despite the fact that it very well might be going at a lot more slow speed will build its energy a lot higher than that of the bike basically constraining the bike the other way. The end-product of estimating an item's position prompts an adjustment of its energy as well as the other way around.

• All Quantum conduct follows this standard and it is significant in deciding otherworldly line widths, as the vulnerability in energy of a framework compares to a line width found in locales of the light range investigated in Spectroscopy.

• Application and utilisation of Heisenberg Uncertainity principle. Assuming Δx is the error in the measurement of position and Δp is the error in the measurement of momentum, then

$\Delta X \times \Delta p \geq h/4\pi.$

• We all know that momentum, p = mv, (momentum=mass ×velocity) Heisenberg's uncertainty principle formula can be also be written as- $\Delta X \times \Delta mv \geq h/4\pi$ or $\Delta X \times \Delta m \times \Delta v \geq h/4\pi$

• Where, ΔV is the error in the measurement of velocity and assuming mass remaining constant during the experiment,

$\Delta X \times \Delta V \geq h/4\pi m$

• Accurate measurement of position or momentum automatically indicates larger uncertainty (error) in the measurement of the other quantity. Applying the Heisenberg principle to an electron in an orbit of an atom, with h(Planck's constant) = $6.626 \times 10\text{-}34$Js and m= $9.11 \times 10\text{-}31$Kg,

$\Delta X \times \Delta V \geq 6.626 \times 10\text{-}34Js/9.11 \times 10\text{-}31$Kg,= $10\text{-}4$ m^{2}S^{-1}

Assuming the place of the electron is estimated precisely to its size (10-10m),, then, at that point, the blunder in the estimation of its speed will be equivalent or bigger than $10^{\wedge}6$m or 1000Km.

Clarifying Heisenberg Uncertainity principle With A Model.

Electromagnetic radiations and tiny matter waves show a double nature of mass/energy and wave character. Position and speed/energy of perceptible matter waves not really settled precisely, at the same time. For instance, the area and speed of a moving vehicle not set in stone simultaneously, with least blunder. Be that as it may, in infinitesimal particles, it won't be imaginable to fix the position and measure the speed/energy of the molecule all the while.

An electron in an atom has a mass of $9.91 \times 10\text{-}31Kg$

Unaided eyes won't see such little particles. A strong light might crash into the electron and enlighten it. Enlightenment helps in recognizing and estimating the place of the electron. The impact of the strong light source, while helping in ID expands the energy of the electron and makes it get away from the underlying position. Along these lines, when fixing the position, the speed/energy of the molecule would have transformed from the first worth. Henceforth, when the position is precise the blunder happens in the estimation of speed or force. Similarly, the estimation of energy precisely will change the position.

Thus, anytime, either position or energy must be estimated precisely.

Synchronous estimation of the two of them will have a mistake in both position and energy. Heisenberg evaluated the blunder in the estimation of both position and force simultaneously.

• Heisenberg's γ-beam Magnifying lens: A striking psychological test outlining the uncertainty principle is Bohr's/Heisenberg's Gamma-beam magnifying instrument. To notice a molecule, say an electron, we focus it with the light beam of frequency λ and gather the Compton dispersed light in a magnifying instrument objective whose distance across subtends a point θ with the electron. The accuracy with which the electron can be found, Delta x, is characterized by the settling force of the magnifying lens, $\sin\theta = \Delta x \lambda \Rightarrow \Delta x = \sin\theta\lambda$

• Apparently by making λ little, to that end we pick γ-beam, and by making sin θ huge,

Delta x can be made as little as wanted. Be that as it may, as indicated by the vulnerability standard, we can do as such just to the detriment of our insight into x-part of electron energy.

• To record the Compton dissipated photon by the magnifying instrument, the photon should remain in the cone of point θ and subsequently its x-part of the energy can change inside ±(h/λ) sin θ. This suggests, the extent of the backlash force of the electron is dubious by

$\Delta px=(2h/\lambda)\sin\theta$

$=(2h/\lambda)\sin\theta$

The result of uncertainty gives.

$\Delta X \; \Delta px=(\lambda/\sin\theta)(2h/\lambda)\sin\theta=4\pi h$

Is Heisenberg's uncertainty Guideline Recognizable in Every Matter Wave?

Heisenberg's standard is applicable to all matter waves. The estimation mistake of any two form properties, whose aspects end up being joule sec, similar to positionforce, time-energy will be directed by Heisenberg's worth.

In any case, it will be perceptible and of importance just for little particles like an electron with extremely low mass. A greater molecule with a weighty mass will demonstrate the blunder to be tiny and irrelevant.

Outcomes

The Heisenberg Standard has huge bearing on rehearsed science and how investigations are planned. Think about estimating the energy or position of a molecule. To make an estimation, a communication with the molecule should happen that will modify it's different factors. For instance, to quantify the place of an electron there should be an impact between the electron and another molecule like a photon. This will bestow a portion of the second molecule's energy onto the electron being estimated and along these lines changing it. A more precise estimation of the electron's position would require a molecule with a more modest frequency, and thusly be more lively, however at that point this would change the force significantly really during impact. An investigation intended to decide energy would similarly affect position. Therefore, examinations can just assemble data about a solitary variable at a time with any measure of precision. Worked

1. The uncertainty in the momentum Δp of a football thrown by Conor Coady during the superbowl traveling at 20m/s is $1 \times 10\text{-}5$ of its momentum. Calculate the uncertainty in its position Δx? Mass=0.50kg

Note that h(Planck's Constant)=$6.626 \times 10\text{-}34$js

Solution Recall.

Momentum(p)=mass(m)×Velocity(V)

P=0.50kg×20m/s

P=10Kgm/s

$\Delta p = p \ (1 \times 10\text{-}5)$

$\Delta p = 10$Kgm/s$(1 \times 10\text{-}5)$

$\Delta p = 10 \times 10\text{-}5$Kgm/s

Since $\Delta X \times \Delta p \geq h/4\pi$.

Therefore, $\Delta X \geq h/(4\pi \times \Delta p)$

$\Delta x \geq (6.626 \times 10\text{-}34js)/(4\pi \times 10 \times 10\text{-}5kgm/s)$

$\Delta x \geq (5.3 \times 10\text{-}31)m$

2.An electron in a molecule of water is traveling at the speed of 30m/s and the uncertainty Δp in its momentum is $2 \times 10\text{-}6$. Calculate its Δx if the mass of an electron is $9.1 \times 10\text{-}31kg$

Solution

Recall.

$Momentum(p) = mass(m) \times Velocity(V)$

$P = 9.1 \times 10\text{-}31kg \times 30m/s$

$P = 2.73 \times 10\text{-}32Kgm/s$

$\Delta p = p\ (2 \times 10\text{-}6)$

$\Delta p = 2.73 \times 10\text{-}32Kgm/s$

$\Delta p = 5.46 \times 10\text{-}38Kgm/s$

Quantum Physics for beginners

Since $\Delta X \times \Delta p \geq h/4\pi$.

Therefore, $\Delta X \geq h/(4\pi \times \Delta p)$

$\Delta x \geq (6.626 \times 10\text{-}34 \text{js})/(4\pi \times 5.46 \times 10\text{-}38 \text{kgm/s})$

$\Delta x \geq 965.7\text{m}$

Chapter Three
Quantum Mechanics

Science managing the conduct of issue and light on the nuclear and subatomic scale. It endeavors to portray and represent the properties of particles and molecules and their constituents—electrons, protons, neutrons, and other more exclusive particles like quarks and gluons. These properties incorporate the cooperations of the particles with each other and with electromagnetic radiation (i.e., light, X-beams, and gamma beams). The conduct of issue and radiation on the nuclear scale regularly appears to be curious, and the results of quantum hypothesis are likewise hard to comprehend and to accept. Its ideas oftentimes struggle with sound judgment thoughts got from perceptions of the regular world. There is not a remotely good excuse, nonetheless, why the conduct of the nuclear world ought to adjust to that of the natural, enormous scope world. It is essential to understand that quantum mechanics is a part of physical science and that the matter of physical science is to depict and represent the way the world—on both the huge and the limited scale—really is and not how one envisions it or would like it to be. Quantum mechanics is the part of physical science identifying with the tiny. It brings about what might give off an impression of being some exceptionally unusual decisions about the actual world. At the size of molecules and electrons, a significant number of the situations of old style mechanics, which depict how things move at regular sizes and velocities, stop to be helpful. In old style mechanics, objects exist in a particular spot at a particular time. In any case, in quantum mechanics, objects rather exist in a dimness of likelihood; they have a specific shot at being at point A, one more opportunity of being at point B, etc.

The investigation of quantum mechanics is remunerating for a long time. In the first place, it delineates the fundamental philosophy of material science. Second, it has been gigantically fruitful in giving right outcomes in basically every circumstance to which it has been applied. There is, be that as it may, a fascinating mystery. Despite the mind-boggling functional achievement of quantum mechanics, the establishments of the subject contain annoying issues—specifically, issues concerning the idea of estimation. A fundamental element of quantum mechanics is that it is by and large outlandish, even on a basic level, to gauge a framework without upsetting it; the itemized idea of this aggravation and the specific place where it happens are dark and dubious. Subsequently, quantum mechanics pulled in probably the ablest researchers of the twentieth century, and they raised what is maybe the best learned structure of the period.

Chronicled premise of quantum hypothesis

Essential contemplations

At a basic level, both radiation and matter have attributes of particles and waves. The progressive acknowledgment by researchers that radiation has molecule like properties and that matter has wavelike properties given the stimulus to the improvement of quantum mechanics. Affected by Newton, most physicists of the eighteenth century accepted that light comprised of particles, which they called corpuscles. From around 1800, proof started to collect for a wave hypothesis of light. At about this time Thomas Young showed that, assuming monochromatic light goes through a couple of cuts, the two arising radiates meddle, so a periphery example of then again brilliant and dim groups shows up on a screen. The groups are promptly clarified by a wave hypothesis of light. As per the hypothesis, a brilliant band is created when the peaks (and box) of the waves from the two cuts show up together at the screen; a dull band is delivered when the peak of one wave shows up simultaneously as the box of the other, and the impacts of the two light pillars drop. Starting in 1815, a progression of analyses by Augustin-Jean Fresnel of France and others showed that, when an equal light emission goes through a solitary cut, the arising bar is presently not equal however begins to wander; this peculiarity is known as diffraction. Given the frequency of the light and the math of the mechanical assembly (i.e., the detachment and widths of the cuts and the separation from the cuts to the screen), one can utilize the wave hypothesis to work out the normal example for each situation; the hypothesis concurs exactly with the test information.

Early turns of events

Planck's radiation law

Before the finish of the nineteenth century, physicists generally acknowledged the wave hypothesis of light. Be that as it may, however the thoughts of traditional material science clarify impedance and diffraction peculiarities identifying with the spread of light, they don't represent the assimilation and discharge of light.

Quantum Physics for beginners

All bodies transmit electromagnetic energy as hotness; truth be told, a body discharges radiation at all frequencies. The energy transmitted at various frequencies is a most extreme at a frequency that relies upon the temperature of the body; the more sizzling the body, the more limited the frequency for greatest radiation. Endeavors to ascertain the energy circulation for the radiation from a blackbody utilizing traditional thoughts were fruitless. (A blackbody is a theoretical ideal body or surface that retains and reemits all brilliant energy falling on it.) One equation, proposed by Wilhelm Wien of Germany, disagreed with perceptions at long frequencies, and another, proposed by Lord Rayleigh (John William

Strutt) of England, contradicted those at short frequencies. In 1900 the German hypothetical physicist Max Planck made an intense idea. He accepted that the radiation energy is produced, not consistently, yet rather in discrete bundles called quanta. The energy E of the quantum is identified with the recurrence ν by $E = h\nu$. The amount h, presently known as Planck's consistent, is a general steady with the surmised worth of $6.62607 \times 10-34$ joule·second. Planck showed that the determined energy range then, at that point, concurred with perception over the whole frequency range.

Three progressive standards

Quantum mechanics (QM) created over numerous many years, starting as a bunch of dubious numerical clarifications of analyses that the math of traditional mechanics couldn't clarify. It started at the turn of the twentieth century, around the very time that Albert Einstein distributed his hypothesis of relativity, a different numerical unrest in material science that portrays the movement of things at high velocities. In contrast to relativity, nonetheless, the starting points of QM can't be ascribed to any one researcher. Rather, numerous researchers added to an establishment of three progressive rules that step by step acquired acknowledgment and trial check somewhere in the range of 1900 and 1930. They are:

Quantized properties: Certain properties, like position, speed and shading, can now and again just happen in explicit, limited sums, similar as a dial that "clicks" from one number to another. This tested a central presumption of traditional mechanics, which said that such properties should exist on a smooth, persistent range. To portray the possibility that a few properties "clicked" like a dial with explicit settings, researchers begat "quantized."

Particles of light: Light can in some cases act as a molecule. This was at first met with cruel analysis, as it negated 200 years of examinations showing that light acted as a wave; similar as waves on the outer layer of a quiet lake. Light acts in basically the same manner in that it skips off dividers and curves around corners, and that the peaks and box of the wave can add up or counterbalance. Added wave peaks bring about more splendid light, while waves that counterbalance produce haziness. A light source can be considered as a ball on a stick being musically dunked in the focal point of a lake. The tone discharged compares to the distance between the peaks, which is dictated by the speed of the ball's cadence.

Rushes of issue: Matter can likewise act as a wave. This opposed the approximately 30 years of investigations showing that matter (like electrons) exists as particles.

Quantized properties?

In 1900, German physicist Max Planck looked to clarify the conveyance of shadings radiated over the range in the shine of super hot and white-hot articles, like light fibers. When understanding the condition he had inferred to depict this dissemination, Planck acknowledged it suggested that mixes of just certain shadings (though an incredible number of them) were discharged, explicitly those that were entire number products of some base worth. Some way or another, colors were quantized! This was sudden in light of the fact that light was perceived to go about as a wave, implying that upsides of shading ought to be a consistent range. What could be denying particles from delivering the shadings between these entire number products? This appeared to be really weird that Planck viewed quantization as just a numerical stunt. As indicated by Helge Kragh in his 2000 article in Physics World magazine, "Max Planck, the Reluctant Revolutionary," "Assuming an upheaval happened in material science in December 1900, no one appeared to see it. Planck was no special case … "

Planck's condition additionally contained a number that would later turn out to be vital to future advancement of QM; today, it's known as "Planck's Constant."

Quantization assisted with clarifying different secrets of physical science. In 1907, Einstein utilized Planck's speculation of quantization to clarify why the temperature of a strong changed by various sums assuming you put a similar measure of hotness into the material yet changed the beginning temperature.

Since the mid 1800s, the study of spectroscopy had shown that various components transmit and retain explicit shades of light called "phantom lines." Though spectroscopy was a solid technique for deciding the components contained in articles, for example, far off stars, researchers were astounded with regards to why every component emitted those particular lines in any case. In 1888, Johannes Rydberg determined a condition that depicted the unearthly lines radiated by hydrogen, however no one could clarify why the condition worked. This changed in 1913 when Niels Bohr applied Planck's speculation of quantization to Ernest Rutherford's 1911 "planetary" model of the iota, which hypothesized that electrons circled the core the same way that planets circle the sun. As indicated by Physics 2000 (a site from the University of Colorado), Bohr recommended that electrons were confined to "extraordinary" circles around a molecule's core. They could "bounce" between extraordinary circles, and the energy delivered by the leap caused explicit shades of light, seen as otherworldly lines. However quantized properties were developed as yet a simple numerical stunt, they clarified such a lot of that they turned into the establishing rule of QM.

Particles of light?

In 1905, Einstein distributed a paper, "Concerning a Heuristic Point of View Toward the Emission and Transformation of Light," wherein he imagined light voyaging not as a wave, but rather as some way of "energy quanta." This parcel of energy, Einstein recommended, could "be assimilated or created uniquely overall," explicitly when a molecule "hops" between quantized vibration rates. This would likewise apply, as would be shown a couple of years after the fact, when an electron "hops" between quantized circles. Under this model, Einstein's "energy quanta" contained the energy distinction of the leap; when partitioned by Planck's consistent, that not really set in stone the shade of light conveyed by those quanta.

With this better approach to imagine light, Einstein offered bits of knowledge into the conduct of nine unique peculiarities, including the particular shadings that Planck depicted being radiated from a light fiber. It likewise clarified how specific shades of light could discharge electrons off metal surfaces, a peculiarity known as the "photoelectric impact." However, Einstein wasn't completely defended in taking this jump, said Stephen Klassen, an academic partner of physical science at the University of Winnipeg. In a 2008 paper, "The Photoelectric Effect: Rehabilitating the Story for the Physics Classroom," Klassen states that Einstein's energy quanta aren't required for clarifying those nine peculiarities. Certain numerical medicines of light as a wave are as yet equipped for depicting both the particular shadings that Planck portrayed being produced from a light fiber and the photoelectric impact. To be sure, in Einstein's disputable winning of the 1921 Nobel Prize, the Nobel council just recognized "his disclosure of the law of the photoelectric impact," which explicitly didn't depend on the idea of energy quanta.

Approximately twenty years later Einstein's paper, the expression "photon" was advocated for portraying energy quanta, because of the 1923 work of Arthur Compton, who showed that light dispersed by an electron shaft changed in shading. This showed that particles of light (photons) were for sure slamming into particles of issue (electrons), consequently affirming Einstein's theory. At this point, obviously light could act both as a wave and a molecule, putting light's "wave-molecule duality" into the establishment of QM.

Floods of issue?

Since the revelation of the electron in 1896, proof that all matter existed as particles was gradually assembling. All things considered, the exhibition of light's wave-molecule duality made researchers question whether matter was restricted to acting just as particles. Maybe wave-molecule duality could sound valid for issue also? The main researcher to gain significant ground with this thinking was a French physicist named Louis de Broglie. In

97

1924, de Broglie utilized the conditions of Einstein's hypothesis of unique relativity to show that particles can display wave-like qualities, and that waves can show molecule like attributes. Then, at that point, in 1925, two researchers, working autonomously and utilizing separate lines of numerical reasoning, applied de Broglie's thinking to clarify how electrons zoomed around in particles (a peculiarity that was unexplainable utilizing the conditions of traditional mechanics). In Germany, physicist Werner Heisenberg (collaborating with Max Born and Pascual Jordan) achieved this by creating "lattice mechanics." Austrian physicist Erwin Schrödinger fostered a comparable hypothesis called "wave mechanics." Schrödinger displayed in 1926 that these two methodologies were same (however Swiss physicist Wolfgang Pauli sent an unpublished outcome to Jordan showing that framework mechanics was more finished).

The Heisenberg-Schrödinger model of the particle, in which every electron goes about as a wave (now and again alluded to as a "cloud") around the core of an iota supplanted the Rutherford-Bohr model. One specification of the new model was that the closures of the wave that frames an electron should meet. In "Quantum Mechanics in Chemistry, third Ed." (W.A. Benjamin, 1981), Melvin Hanna states, "The burden of the limit conditions has confined the energy to discrete qualities." An outcome of this specification is that main entire quantities of peaks and box are permitted, which clarifies why a few properties are quantized. In the Heisenberg-Schrödinger model of the particle, electrons obey a "wave work" and possess "orbitals" rather than circles. Not at all like the roundabout circles of the Rutherford-Bohr model, nuclear orbitals have an assortment of shapes going from circles to free weights to daisies.

In 1927, Walter Heitler and Fritz London further created wave mechanics to show how nuclear orbitals could consolidate to frame sub-atomic orbitals, successfully showing why particles attach to each other to shape particles. This was one more issue that had been unsolvable utilizing the math of old style mechanics. These bits of knowledge led to the field of "quantum science."

The vulnerability standard

Additionally in 1927, Heisenberg made one more significant commitment to quantum material science. He contemplated that since issue goes about as waves, a few properties, like an electron's position and speed, are "correlative," which means there's a limit (identified with Planck's steady) to how well the accuracy of every property can be known. Under what might come to be designated "Heisenberg's vulnerability guideline," it was contemplated that the more exactly an electron's position is known, the less definitively its speed can be known, as well as the other way around. This vulnerability rule applies to ordinary size objects also, however isn't perceptible on the grounds that the absence of accuracy is uncommonly little. As per Dave Slaven of Morningside College (Sioux City, IA), assuming a baseball's speed is known to inside an accuracy of 0.1 mph, the most extreme accuracy to which it is feasible to realize the ball's position is 0.00000000000000000000000000008 millimeters.

Forward

The standards of quantization, wave-molecule duality and the vulnerability guideline introduced another period for QM. In 1927, Paul Dirac applied a quantum comprehension of electric and attractive fields to lead to the investigation of "quantum field hypothesis" (QFT), which treated particles (like photons and electrons) as invigorated conditions of a basic actual field. Work in QFT proceeded for 10 years until researchers hit a barrier: Many conditions in QFT quit seeming to be OK since they created consequences of vastness. Following a time of stagnation, Hans Bethe made a forward leap in 1947 utilizing a method called "renormalization." Here, Bethe understood that all limitless outcomes identified with two peculiarities (explicitly "electron self-energy" and "vacuum polarization") to such an extent that the noticed upsides of electron mass and electron charge could be utilized to make every one of the vast qualities vanish.

Since the forward leap of renormalization, QFT has filled in as the establishment for creating quantum hypotheses about the four major powers of nature: 1) electromagnetism, 2) the feeble atomic power, 3) the solid atomic power and 4) gravity. The principal understanding given by QFT was a quantum portrayal of electromagnetism through "quantum electrodynamics" (QED), which made progress in the last part of the 1940s and mid 1950s. Next was a quantum portrayal of the frail atomic power, which was bound together with electromagnetism to fabricate "electroweak hypothesis" (EWT) all through the 1960s. At long last came a quantum treatment of the solid atomic power utilizing "quantum chromodynamics" (QCD) during the 1960s and 1970s. The speculations of QED, EWT and QCD together structure the premise of the Standard Model of molecule material science. Sadly, QFT presently can't seem to create a quantum hypothesis of gravity. That journey proceeds with today in the investigations of string hypothesis and circle quantum gravity.

Quantum electrodynamics

The utilization of quantum hypothesis to the collaboration among electrons and radiation requires a quantum therapy of Maxwell's field conditions, which are the establishments of electromagnetism, and the relativistic hypothesis of the electron detailed by Dirac (see above Electron twist and antiparticles). The subsequent quantum field hypothesis is known as quantum electrodynamics, or QED.

QED represents the conduct and associations of electrons, positrons, and photons. It manages processes including the production of material particles from electromagnetic energy and with the opposite processes wherein a material molecule and its antiparticle obliterate one another and produce energy. At first the hypothesis was plagued with considerable numerical challenges, on the grounds that the determined upsides of amounts, for example, the charge and mass of the electron ended up being boundless. Notwithstanding, a clever arrangement of procedures created (in the last part of the 1940s) by Hans Bethe, Julian S. Schwinger, Tomonaga Shin'ichirō, Richard P. Feynman, and others managed the vast qualities to get limited upsides of the actual amounts. Their technique is known as renormalization. The hypothesis has given some surprisingly precise forecasts.

As indicated by the Dirac hypothesis, two specific states in hydrogen with various quantum numbers have a similar energy. QED, nonetheless, predicts a little contrast in their energies; the distinction might be controlled by estimating the recurrence of the electromagnetic radiation that produces changes between the two states. This impact was first estimated by

Willis E. Sheep, Jr., and Robert Retherford in 1947. Its actual beginning lies in the connection of the electron with the irregular changes in the encompassing electromagnetic field. These variances, which exist even without a trace of an applied field, are a quantum peculiarity. The precision of investigation and hypothesis in this space might be measured by two late qualities for the partition of the two states, communicated as far as the recurrence of the radiation that creates the advances:

experiment (1982)   1,057,858 ± 2 kilohertz
theory (1975)   1,057,864 ±14 kilohertz.

A significantly more tremendous illustration of the accomplishment of QED is given by the worth to μe, the attractive dipole snapshot of the free electron. Since the electron is turning and has electric charge, it acts like a little magnet, the strength of which is communicated by the worth of μe. As indicated by the Dirac hypothesis, μe is by and large equivalent to μB = $e\hbar/2me$, an amount known as the Bohr magneton; nonetheless, QED predicts that μe = $(1 + a)\mu B$, where a will be a modest number, roughly $1/860$. Once more, the actual beginning of the QED amendment is the cooperation of the electron with arbitrary motions in the encompassing electromagnetic field. The best trial assurance of μe includes estimating not simply the amount but rather the little revision term $\mu e - \mu B$. This enormously improves the affectability of the analysis. The latest outcomes

for the worth of an are

experiment (1984)	$(115,965,219 \pm 1) \times 10^{-11}$
theory (1986)	$(115,965,227 \pm 10) \times 10^{-11}$.

Since an itself addresses a little rectification

term, the attractive dipole snapshot of the

electron is estimated with a precision of around

one section in 1011. One of the not really settled

amounts in physical science, the attractive

dipole snapshot of the electron can be

determined accurately from quantum

hypothesis to inside around one section in 1010.

The understanding of quantum mechanics

Despite the fact that quantum mechanics has

been applied to issues in physical science with

extraordinary achievement, a portion of its

thoughts appear to be abnormal. A couple of their suggestions are thought of as here.

The electron: wave or molecule?

Youthful's previously mentioned test in which an equal light emission light is gone through a couple of thin equal cuts (Figure 5A) has an electron partner. In Young's unique trial, the force of the light fluctuates with course subsequent to going through the cuts (Figure below). The power sways on account of obstruction between the light waves arising out of the two cuts, the pace of wavering relying upon the frequency of the light and the division of the cuts. The wavering makes a periphery example of rotating light and dull groups that is balanced by the diffraction design from each cut. In the event that one of the cuts is covered, the impedance borders vanish, and just the diffraction design (displayed as a wrecked line in Figure below) is noticed.

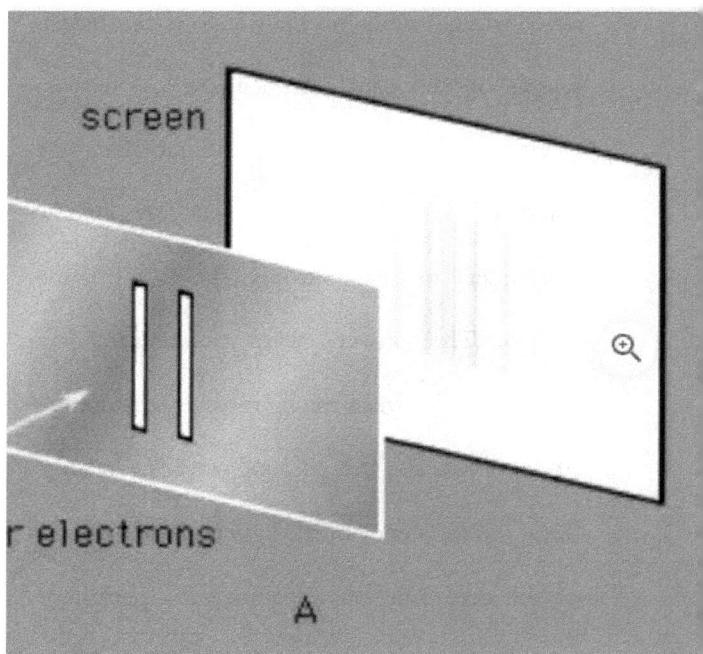

screen

r electrons

A

Youthful's examination can be rehashed with electrons all with a similar energy. The screen in the optical trial is supplanted by a firmly separated matrix of electron finders. There are numerous gadgets for distinguishing electrons; the most well-known are scintillators. At the point when an electron goes through a shining material, for example, sodium iodide, the material delivers a light glimmer which gives a voltage beat that can be enhanced and recorded. The example of electrons recorded by every indicator is equivalent to that anticipated for waves with frequencies given by the de Broglie recipe. Consequently, the trial gives convincing proof to the wave conduct of electrons.

Assuming the trial is rehashed with an extremely feeble wellspring of electrons so just a single electron goes through the cuts, a solitary locator enlists the appearance of an electron. This is a very much restricted occasion normal for a molecule. Each time the examination is rehashed, one electron goes through the cuts and is identified. A diagram plotted with finder position along one hub and the quantity of electrons along different looks precisely like the swaying impedance design in Figure above. Along these lines, the force work in the figure is corresponding to the likelihood of the electron moving a specific way later it has gone through the cuts. Aside from its units, the capacity is indistinguishable from $\Psi 2$, where Ψ is the arrangement of the time-autonomous Schrödinger condition for this specific examination.

Assuming that one of the cuts is covered, the periphery design vanishes and is swapped by the diffraction design for a solitary cut. Along these lines, the two cuts are expected to deliver the periphery design. In any case, assuming the electron is a molecule, it appears to be sensible to assume that it went through just one of the cuts. The mechanical assembly can be changed to find out what cut by putting a dainty wire circle around each cut. At the point when an electron goes through a circle, it creates a little electric sign, showing what cut it went through. In any case, the impedance periphery design then, at that point, vanishes, and the single-cut diffraction design returns. Since the two cuts are required for the impedance example to show up and since it is difficult to realize what cut the electron went through without annihilating that example, one is compelled to the end that the electron goes through the two cuts simultaneously.

In outline, the test shows both the wave and molecule properties of the electron. The wave property predicts the likelihood of bearing of movement before the electron is identified; then again, the way that the electron is distinguished in a specific spot shows that it has molecule properties. In this manner, the response to the inquiry whether the electron is a wave or a molecule is that it is not one or the other. It is an item showing either wave or molecule properties, contingent upon the kind of estimation that is made on it. At the end of the day, one can't discuss the inherent properties of an electron; all things being equal, one should think about the properties of the electron and estimating mechanical assembly together.

Chapter Four
Schrödinger Wave Equation

What is the Schrodinger Equation?

The Schrödinger condition (otherwise called Schrödinger's wave condition) is an incomplete differential condition that portrays the elements of quantum mechanical frameworks through the wave work. The direction, the situating, and the energy of these frameworks can be recovered by settling the Schrödinger condition. All of the data for a subatomic molecule is encoded inside a wave work. The wave capacity will fulfill and can be addressed by utilizing the Schrodinger condition. The Schrodinger condition is one of the major sayings that are presented in undergrad material science. It is additionally progressively normal to observe the Schrödinger condition being presented inside the electrical designing schedule in colleges as it is material to semiconductors. Tragically, it is just expressed as a propose in the two cases and never determined in any significant manner. This is very disappointing as almost all the other things educated in undergrad quantum material science is based upon this establishment. In this article, we will get the condition without any preparation and I'll put forth a valiant effort to show each progression taken. Curiously, the contentions we will make are simply equivalent to those taken by Schrödinger himself so you can see the lines of reasoning a goliath was making in his time. As an update, here is the time-subordinate Schrödinger condition in 3-

aspects (for a non-relativistic molecule) in the
entirety of its excellence:

$$i\hbar\frac{\partial}{\partial t}\Psi(\vec{r},t) = \left[\frac{-\hbar^2}{2m}\nabla^2 + V(\vec{r},t)\right]\Psi$$

The Schrodinger Equation

The Schrödinger condition, now and again
called the Schrödinger wave condition, is an
incomplete differential condition. It utilizes the
idea of energy protection (Kinetic Energy +
Potential Energy = Total Energy) to acquire
data about the conduct of an electron bound to
a core. It does this by permitting an electron's
wave work, Ψ, to be determined. Addressing the
Schrödinger condition gives us Ψ and $\Psi2$. With
these we get the quantum numbers and the

shapes and directions of orbitals that describe electrons in a particle or atom. The Schrödinger condition gives definite arrangements just for cores with one electron: H, He+, Li2+, Be3+, B4+, C5+, and so forth In numerical language, we say that logical answers for Ψ are conceivable just for one-electron frameworks. One-electron frameworks are regularly portrayed as hydrogenic - signifying "like hydrogen." For any remaining iotas, particles, and atoms, no scientific answers for Ψ are conceivable; estimation strategies for computation, like the orbital guess and variety hypothesis, are then used. There is a period subordinate Schrödinger condition and a period free Schrödinger condition. The time-free condition considers the electron's quantum state to be perpetual, subsequently it thinks about the electron as a standing wave. The time-free condition permits electron densities (for example the sizes and states of nuclear and sub-atomic orbitals) to be observed utilizing $\Psi 2$, the square of the wave work.

The p orbitals beneath are instances of $\Psi 2$:

$$p_x \qquad\qquad p_y$$

The Time-Independent Schrödinger Equation

The time-autonomous Schrödinger condition can be communicated in exceptionally packed numerical shorthand as:

$$\hat{H} \, \Psi = E \, \Psi$$

Hamiltonian
Operator
(Energy operator)

Energy
eigenvalue

The condition applies to electrons going at non-relativistic velocities. (This implies it requires changes before it tends to be applied high mass components.) The condition says:

The amount of the Wave Function's Kinetic Energy and Potential Energy = The Wave Function's Total Energy

The time-autonomous condition can be written in any reasonable direction framework, for example, Cartesian directions (x,y,z). For hydrogenic iotas, circular polar directions are more reasonable, henceforth:

$$\frac{-\hbar^2}{2m}\nabla^2\Psi(r) + V(r)\Psi(r) = E\Psi$$

$$\underbrace{Kinetic\ Energy} + \underbrace{Potential\ Energy} = \underbrace{Tot\ Ene}$$

schrodinger-condition time-ind-laplacian

h is the diminished Planck steady,

m is the electron mass,

is the Laplacian administrator,

Ψ is the wave work,

V is the possible energy,

E is the energy eigenvalue,

(r) indicates the amounts are elements of circular polar directions (r, θ, φ)

Requirements Result in Quantization

The condition is tackled to track down Ψ Limitations set on addressing the condition produce quantization - for example the arrangements found for Ψ are confined to specific qualities and any remaining qualities are illegal.

These imperatives are:

•   Ψ furthermore its first fractional subsidiary should be nonstop.

•   Ψ should be normalizable: this implies there is a 100% likelihood of the electron being some place in the universe. For a genuinely esteemed Ψ, standardization requires that:

$$\int_{-\infty}^{\infty} \Psi^2 \, dr = 1$$

As the separation from the core builds, the electron becomes confined and is not generally bound. for example as r → ∞, H → H+ + e-.

The Wave Function Produces Quantum Numbers

The electron's wavefunction exists in three aspects, accordingly arrangements of the Schrödinger condition have three sections. These are gotten expressly by a technique for addressing halfway differential conditions called isolating the factors. Doing this, we get:

$$\Psi(r, \theta, \varphi) = R(r)\, P(\theta)\, F(\varphi)$$

It just so happens, answers for Ψ are just conceivable when:

- In $R(r)$, a consistent, call it n, has values 1, 2, 3, 4,....

- In $P(\theta)$, a consistent, call it l, has values 0, 1, 2, 3,...(n-1)

- In F(φ), a consistent, call it ml, has values - l, (l+1),...0..., (l+1), l

Thus from the wave work Ψ the Schrödinger condition has conveyed the three quantum numbers that portray electron conduct in a particle.

• n: the main quantum number

• l: the orbital rakish force quantum number

• ml: the attractive quantum number

The Wave Function Produces Orbital Shapes and Sizes

$\Psi 2$, the likelihood thickness, characterizes the shapes and sizes of electron orbitals.

Chapter Five
String Theory

The present physicists are battling with a bind. They have acknowledged two separate speculations that clarify how the universe functions: Albert Einstein's overall theory of relativity, which portrays the universe on an extremely enormous scope, and quantum mechanics, which depicts the universe for a tiny scope. Both of these speculations have been upheld predominantly by test proof. Sadly, these theories don't supplement each other. General relativity, which portrays how gravity works, suggests a smooth and streaming universe of twists and bends in the texture of spacetime. Quantum mechanics—with its vulnerability rule—infers that on an imperceptibly limited scale, the universe is a violent, turbulent spot where occasions must be anticipated with probabilities. In two situations where the contending hypotheses must both be applied— to portray the big bang theory and the profundities of black holes—the conditions separate.

Quantum Physics for beginners

Most physicists struggle tolerating that the universe works as per two independent (and once in a while problematic) speculations. They think almost certainly, the universe is represented by a solitary hypothesis that clarifies all perceptions and information. Therefore, Physicists are on the chase after a brought together hypothesis. Such a hypothesis would unite under one umbrella every one of the four powers of nature: gravity, the most fragile of the four, as clarified by broad relativity; and electromagnetism and the solid and feeble powers, as clarified by quantum field hypothesis. Einstein sought after a bound together hypothesis by attempting to join electromagnetism and gravity.

Superstring theory, likewise called string theory, is the current definition of this continuous mission.

Definition and Explanation of String theory

String theory is a hypothetical structure where the point-like particles of molecule physical science are supplanted by one-layered articles called strings. String hypothesis portrays how these strings engender through space and connect with one another. String theory, in particle physics,is one which endeavors to consolidate quantum mechanics with Albert Einstein's overall hypothesis of relativity. The name string theory comes from the displaying of subatomic particles as small one-layered "stringlike" substances rather than the more regular methodology wherein they are demonstrated as zero-layered point particles. The hypothesis imagines that a string going through a specific method of vibration relates to a molecule with unmistakable properties like mass and charge. During the 1980s, physicists understood that string theory could consolidate each of the four of nature's powers—gravity, electromagnetism, solid power, and feeble power—and a wide range of issue in a solitary quantum mechanical system, recommending that it very well may be the since quite a while ago looked for bound together field hypothesis. While string hypothesis is as yet a dynamic space of exploration that is going through quick turn of events, it remains principally a numerical build since it presently can't seem to connect with trial perceptions.

String theory, Quantum theory and Einstein's theory of Relativity.

In 1905 Einstein bound together space and time with his extraordinary hypothesis of relativity, showing that movement through space influences the progression of time. In 1915 Einstein further bound together space, time, and attractive energy with his overall hypothesis of relativity, showing that twists and bends in reality are liable for the power of gravity. These were stupendous accomplishments, yet Einstein longed for a much more terrific unification. He imagined one strong system that would

135

represent space, time, and nature's powers in general— something he called a brought together hypothesis. Throughout the previous thirty years of his life, Einstein perseveringly sought after this vision. In spite of the fact that now and again bits of gossip spread that he had succeeded, closer investigation consistently ran such expectations. The majority of Einstein's counterparts viewed as the quest for a bound together hypothesis to be a sad, if not off track, journey.

Interestingly, the essential worry of hypothetical physicists from the 1920s forward was quantum mechanics—the arising system for depicting nuclear and subatomic cycles. Particles at these scales have such little masses that gravity is basically immaterial in their cooperations, thus for a really long time quantum mechanical estimations by and large overlooked general relativistic impacts. All things being equal, by the last part of the 1960s the attention was on an alternate power—the solid power, which ties together the protons and neutrons inside nuclear cores. Gabriele Veneziano, a youthful scholar working at the European Association for Atomic

Exploration (CERN), contributed a vital forward leap in 1968 with his acknowledgment that a 200-year-old recipe, the Euler beta capacity, was fit for clarifying a significant part of the information on the solid power then, at that point, being gathered at different molecule gas pedals all over the planet. A couple of years after the fact, three physicists—Leonard Susskind of Stanford College, Holger Nielsen of the Niels Bohr Establishment, and Yoichiro Nambu of the College of Chicago—fundamentally intensified Veneziano's knowledge by showing that the mathematics underlying his proposition portrayed the vibrational movement of minute fibers of energy that look like small strands of string, moving the name string hypothesis. Generally talking, the hypothesis proposed that the solid power added up to strings tying together particles appended to the strings' endpoints.

String hypothesis was an instinctively alluring proposition, however by the mid-1970s morerefined estimations of the solid power had veered off from its forecasts, driving most scientists to infer that string hypothesis had no significance to the actual universe, regardless of how rich the numerical hypothesis. By and by, few physicists kept on seeking after string hypothesis. Albeit nobody had prevailed with regards to consolidating general relativity and quantum mechanics, fundamental work had set up that such an association would require exactly the massless molecule anticipated by string hypothesis. A couple of physicists contended that string hypothesis, by having this molecule incorporated into its key design, had joined the laws of the huge (general relativity) and the laws of the little (quantum mechanics).

Rather than simply being a depiction of the solid power, these physicists battled, string hypothesis required reevaluation as a basic advance toward Einstein's bound together hypothesis.

The declaration was generally overlooked. String hypothesis had effectively fizzled in its first manifestation as a depiction of the solid power, and many felt it was impossible that it would now win as the answer for a significantly more troublesome issue. This view was supported by string hypothesis' experiencing its own hypothetical issues. As far as one might be concerned, a portion of its situations gave indications of being conflicting; for another, the arithmetic of the hypothesis requested the universe have not recently the three spatial components of normal experience however six others (for an aggregate of nine spatial aspects, or a sum of ten spacetime aspects). By the mid-1990s these and different hindrances were again dissolving the positions of string scholars. Yet, in 1995 another advancement revitalized the field. Edward Witten of the Establishment for Cutting edge Study, expanding on commitments of numerous different physicists, proposed another arrangement of methods that refined the inexact conditions on which all work in string hypothesis had up until this point been based.

These procedures uncovered various new highlights of string hypothesis, including the acknowledgment that the hypothesis has not six but rather seven extra spatial aspects. The more precise conditions additionally uncovered fixings in string hypothesis other than strings— membranelike objects of different aspects, all things considered called branes. At long last, the new methods set up that different variants of string hypothesis created throughout the previous many years were basically no different either way. Scholars call this unification of previously particular string speculations by another name, M-hypothesis, with the significance of M being conceded until the hypothesis is all the more completely perceived.

One more development in string hypothesis occurred in 1997 when Juan Maldacena of Harvard College found the counter de Sitter/conformal field hypothesis (Advertisements/CFT) correspondence. Maldacena observed that a string hypothesis working with a specific climate (including a space-time known as an enemy of de Sitter space) was comparable to a kind of quantum field hypothesis working in a climate with one less spatial aspect. This has ended up being one of the most significant revelations in string hypothesis, building up a strong connection to the more traditional strategies for quantum field hypothesis, giving a precise numerical detailing of string hypothesis in specific conditions, and moving a large number of additional specialized examinations.

Today the comprehension of numerous aspects of string hypothesis is as yet in its developmental stage. Scientists perceive that, albeit wonderful headway has been made in the course of recent many years, on the whole the work is troubled by its piecemeal turn of events, with steady disclosures having been joined like bits of a jigsaw puzzle. That the pieces fit lucidly is noteworthy, yet the bigger picture they are finishing up—the central guideline fundamental the hypothesis—stays puzzling. Similarly squeezing, the hypothesis presently can't seem to be upheld by perceptions and henceforth stays an absolutely hypothetical build. String theory endeavors to bind together every one of the four powers, and in this manner, bring together broad relativity and quantum mechanics. At its center is a genuinely straightforward thought—all particles are made of minuscule vibrating strands of energy. (String hypothesis gets its name from the string-like appearance of these energy strands.) Not at all like ordinary strings, these strings have length (averaging around 10^{-33} centimeters) yet no thickness. String hypothesis infers that the particles that contain all the matter that you find known to mankind— and every one of the powers that permit make a difference to communicate—are made of little vibrating strands of energy.

The presently acknowledged and tentatively checked hypothesis of how the universe chips away at subatomic scales holds that all matter is made out of—and cooperates through—point particles. Known as the Standard Model, this hypothesis portrays the rudimentary particles and three of the four central powers that fill in as the structure blocks for our reality. In string hypothesis, each sort of rudimentary matter molecule—and each kind of central power transporter molecule that intervenes cooperations between matter particles—relates to an interesting string vibrational example, to some degree as various notes played by a violin compare to remarkable string vibrations. How a string vibrates decides the properties—like charge, mass, and twist—of the molecule it is. The conditions of string hypothesis could lead to rudimentary particles like those at present known (electrons, quarks, photons, and so forth), but since definite mathematical forecasts can't yet be made, it is hard to tell whether the collection of conceivable vibrational examples accurately represents all known matter and power transporter particles. Strings can either be open-finished or shut to shape a circle. Regardless of whether a string is open or shut decides the kind of cooperations it can go through.

It is the idea of strings that brings together broad relativity and quantum mechanics. Under

147

quantum field hypothesis, particles connect north of zero distance in spacetime. Under the overall hypothesis of relativity, the speculated power transporter molecule for gravity, the graviton, can't work at zero distance. Strings assist with addressing this situation. Since they are one-layered and have length, they "smear" collaborations over little distances. This spreading smooths out spacetime enough for the graviton to interface with other quantum field particles, consequently binding together the two arrangements of laws. String hypothesis, for all its class, accompanies a cost. For the hypothesis to be steady, the universe should have multiple spatial aspects. Indeed, string hypothesis predicts a universe with nine spatial and once aspect, for a sum of 10 aspects. (The most current form of string hypothesis predicts 11 aspects.) The nine spatial aspects comprise of the three expanded aspects that we experience in daily existence, in addition to six speculated minuscule, nestled into that shouldn't be visible with existing innovations. These additional six aspects happen at each point in the natural three-layered world. The presence of multiple spatial aspects is such a troublesome idea to get a handle on that in any event, string scholars can't envision it. They regularly use analogies to assist with envisioning these reflections.

For instance, picture a piece of paper with a two-layered, level surface. On the off chance that you roll up this surface, it will frame a cylinder, and one aspect will become twisted. Presently envision that you keep moving the surface until it is rolled firmly to the point that the inside nestled into appears to vanish and the cylinder essentially resembles a line. Along these lines, the additional aspects anticipated by string hypothesis are really firmly twisted that they appear to vanish in ordinary experience. These nestled into may take on specific complex designs known as Calabi-Yau shapes. Tragically, a huge number of varieties of these shapes exist, and it is hard to tell which ones may accurately address the additional components of our universe. It is critical to know which ones are right since it is the state of these additional aspects that decides the examples of the string vibrations. These examples, thusly, address every one of the parts that permit the known universe to exist.

These additional aspects may be pretty much as little as 10^{-35} meters or as large as a 10th of a millimeter. On the other hand, the additional aspects could be as huge or bigger than our own universe. Assuming that is the situation, a few physicists accept gravity may be spilling across these additional aspects, which could assist with clarifying why gravity is so frail contrasted with the other three forces. String hypothesis additionally requires each referred to issue molecule to have an at this point unseen comparing "super" power transporter molecule and each known power transporter molecule to have an at this point unseen relating "super" matter molecule. This thought, known as supersymmetry, builds up a connection between matter particles and power transporter particles Called superpartners. These estimated particles are believed to be more huge than their known partners, which might be the reason they have not yet been seen with current molecule gas pedals and identifiers.

At the point when everything began

String hypothesis isn't altogether new. It has been advancing since the last part of the 1960s. At a certain point, there were five varieties of the hypothesis. Then, at that point, during the 1990s a hypothesis known as M-hypothesis arose that brought together the five speculations. Mhypothesis is viewed as the most recent advance in string hypothesis development.

No piece of string hypothesis has been tentatively affirmed. This is to some degree since theoreticians don't yet comprehend the hypothesis alright to make conclusive testable forecasts. Furthermore, strings are believed to be so little—under a billionth of a billionth of the size of an iota—that advancements, for example, current gas pedals and locators aren't adequately strong to distinguish them. While string hypothesis can't yet be tentatively checked, physicists trust that a portion of its aspects can be upheld by conditional proof, for example, exhibiting the presence of:

• Additional dimensions: Physicists trust that current or future molecule gas pedals will actually want to assist with showing the presence of additional dimensions. Identifiers may gauge the missing energy that would have spilled from our dimensions into those additional dimensions, potentially giving proof that these dimensions exist.

• Superpartner particles: Specialists will utilize flow and cutting edge molecule gas pedals to look for the superpartner particles anticipated by string hypothesis.

• Vacillations in foundation radiation: The universe is penetrated by uniform radiation of the exceptionally low temperature of 2.7 degrees Kelvin. This is accepted to be left over from the first extremely high temperature of the enormous detonation. Contrasting the temperatures from various areas in the sky somewhere around 1 degree separated, tiny contrasts in temperature have been found (on the request for 100 thousandth of a degree Kelvin). Researchers are searching for considerably more modest contrasts in temperature of a particular structure that might be left over from the earliest nature of big bang, when the energies expected to make strings might have been achieved.

The potential for what string hypothesis could help clarify is immense.

• It could uncover what occurred when the universe started. The theory of the universe's origin just depicts what occurred after the first tiny part of a second. Under ordinary speculations, preceding that,the universe shrank to zero size—an inconceivability. Under the support of string hypothesis, the universe may never have contracted to a place where it vanished yet rather may have started at a miniscule size—the size of a solitary string.

• String hypothesis could likewise assist with uncovering the idea of black holes, which, while anticipated by broad relativity, have never been completely clarified at the quantum level. Utilizing one sort of string hypothesis, physicists have numerically depicted smaller than usual massless black holes that—in the wake of going through changes in the calculation of string hypothesis' additional aspects—return as rudimentary particles with mass and charge. A few scholars currently believe that black holes and essential particles are indistinguishable and that their apparent contrasts reflect something likened to stage advances, similar to fluid water changing into ice.

• String hypothesis likewise makes the way for various theories about the advancement and nature of reality, for example, how the universe may have looked before the huge explosion or the capacity of room to tear and fix itself or to go through topological changes.

Five basic elements of String theory and their applications.

Five key thoughts are at the core of string hypothesis. Come out as comfortable with these critical components of string hypothesis first thing. They are recorded underneath:

1.   **Strings and membranes**: At the point when the hypothesis was initially evolved during the 1970s, the fibers of energy in string hypothesis were viewed as 1-layered items: strings. (One-layered shows that a string has just one aspect, length, rather than say a square, which has both length and tallness aspects.) These strings came in two structures — closed strings and open strings. An open string has ends that don't contact one another, while a

closed string is a circle with no open end. It was at last observed that these early strings, called Type I strings, could go through five essential sorts of associations. Type I strings can go through five central cooperations, in view of various methods of joining and splitting. The connections depend on a string's capacity to have ends joined and separated. Since the finishes of open strings can consolidate to frame closed strings, you can't develop a string hypothesis without closed strings. This ended up being significant, in light of the fact that the closed strings have properties that cause physicists to accept they may portray gravity. Rather than simply being a hypothesis of issue particles, physicists started to understand that string hypothesis may have the option to clarify gravity and the conduct of particles. Throughout the long term, it was found that the hypothesis required articles other than strings. These articles should be visible as sheets, or branes. Strings can append at one or the two closures to these branes. A 2-layered brane (called a 2-brane).

2. **Quantum gravity:** Present day physics has two essential logical laws: quantum physical science and general relativity. These two logical laws address fundamentally various fields of study. Quantum physics concentrates on the exceptionally littlest items in nature, while relativity will in general concentrate on nature on the size of planets, worlds, and the universe overall. (Clearly, gravity influences little particles as well, and relativity represents this too.) Speculations that endeavor to bind together the two hypotheses are speculations of quantum gravity, and the most encouraging of all such speculations today is string theory.

3. **Unification of forces**: Strongly connected with the subject of quantum gravity, string hypothesis endeavors to bring together the four powers known to man — electromagnetic power, the solid atomic power, the frail atomic power, and gravity — together into one brought together hypothesis. In our universe, these basic powers show up as four unique peculiarities, yet string scholars trust that in the early universe (when there were extraordinarily high energy levels) these powers are completely depicted by strings collaborating with one another..

4.   Supersymmetry: One fundamental nature of string hypothesis is known as supersymmetry, a numerical property that requires each known molecule species to have an accomplice molecule animal categories, called a superpartner. (This property represents string hypothesis frequently being alluded to as superstring hypothesis). Particles in the universe can be partitioned into two kinds: bosons and fermions. String hypothesis predicts that a kind of association, called supersymmetry, exists between these two molecule types. Under supersymmetry, a fermion should exist for each boson as well as the other way around. Tragically, tests have not yet recognized these additional particles. At this point, no superpartner particles have been identified tentatively, yet analysts accept this might be because of their weight: they are heavier than their known partners and require a machine essentially as strong as the Enormous Hadron Collider at CERN to deliver them. If the superpartner particles are found, string hypothesis actually won't be demonstrated right, since more-customary point-molecule speculations have likewise effectively joined supersymmetry into their numerical design. Notwithstanding, the disclosure of supersymmetry would affirm a fundamental component of string hypothesis and give conditional proof that this way to deal with unification is doing great. Supersymmetry is a particular numerical connection between specific components of physical science conditions. It was found outside of string hypothesis, in spite of the fact that its fuse into string hypothesis changed the hypothesis into supersymmetric string hypothesis (or superstring hypothesis) during the 1970s. Supersymmetry inconceivably works on string hypothesis' conditions by permitting specific terms to counteract. Without supersymmetry, the conditions bring about actual irregularities, for example, limitless qualities and fanciful energy levels. Since researchers haven't noticed the particles anticipated by supersymmetry, this is as yet a hypothetical suspicion. Numerous physicists accept that the explanation nobody has noticed the particles is on the grounds that it takes a ton of energy to create them. (Energy is identified with mass by Einstein's popular $E = mc^2$ equation, so it takes energy to make a molecule.) They might have existed in the early universe, however as the universe chilled and energy spread out later the huge explosion, these particles would have imploded into the lower-energy expresses that we notice today. (We may not consider our present universe especially low energy, however contrasted with the exceptional fieriness of the initial couple of seconds later the enormous detonation, it absolutely is.)

Researchers trust that galactic perceptions or tests with particle accelerators will reveal a portion of these higher-energy supersymmetric particles, offering help for this expectation of string theory. Regardless of whether these accelerator based tests are uncertain, there is another way that string hypothesis may one day be tried. Through its effect on the soonest, most outrageous snapshots of the universe, the physical science of string hypothesis might have left weak cosmological marks—for instance, as gravitational waves or a specific example of temperature varieties in the inestimable microwave foundation radiation—that might be perceptible by the up and coming age of accuracy satellite-borne telescopes and identifiers. It would be a fitting end to Einstein's mission for unification assuming a hypothesis of the littlest minute part of issue were affirmed through perceptions of the biggest galactic domains of the universe.

5. **Additional dimension:** One more numerical consequence of string hypothesis is that the hypothesis just seems Okay in a world with in excess of three space apects ! (Our universe has three components of room — left/right, up/down, and front/back.) Two potential clarifications at present exist for the area of the additional aspects: The additional room aspects (by and large six of them) are nestled into, (in string theory) to minuscule sizes, so we never see them. We are stuck on a

3-layered brane, and the additional aspects reach out off of it and are distant to us. A significant space of examination among string scholars is on numerical models of how these additional aspects could be identified with our own. A portion of these new outcomes have anticipated that researchers may before long have the option to recognize these additional aspects (assuming they exist) in forthcoming investigations, since they might be bigger than recently anticipated.

Chapter Six

Triumph Of Quantum

Physics

Quantum physical science has been exceptionally effective in clarifying numerous actual peculiarities, for example, wave-molecule duality. Truth be told, quantum material science was made to clarify actual estimations that old style physical science couldn't clarify. This part is about some of the triumphs of quantum physical science.

The Photoelectric Effect

One more establishing mainstay of quantum physical science was clarifying the photoelectric impact, wherein experimenters focused light on a metal. Regardless of how strong the light, the energy of electrons emitted from the metal didn't rise. Incidentally, the energy of electrons goes up with the frequency of the light, not its force — which gives backing to the light as a surge of discrete photons hypothesis..

Photoelectric effect is a peculiarity wherein electrically charged particles are set free from or inside a material when it retains electromagnetic radiation. The impact is regularly characterized as the launch of electrons from a metal plate when light falls on it. In a more extensive definition, the brilliant energy might be infrared, noticeable, or bright light, X-beams, or gamma beams; the material might be a strong, fluid, or gas; and the delivered particles might be particles (electrically charged iotas or atoms) just as electrons. The peculiarity was in a general sense critical in the advancement of present day material science due to the baffling inquiries it raised with regards to the idea of light—molecule versus wavelike conduct—that were at last settled by Albert Einstein in 1905. The impact stays significant for research in regions from materials science to astronomy, just as framing the reason for an assortment of valuable gadgets.

The photoelectric impact was found in 1887 by the German physicist Heinrich Rudolf Hertz. Regarding work on radio waves, Hertz saw that, when bright light radiates on two metal cathodes with a voltage applied across them, the light changes the voltage at which sparking occurrs. This connection among light and electricity (thus photoelectric) was explained in 1902 by another German physicist, Philipp Lenard. He exhibited that electrically charged particles are freed from a metal surface when light shine on it and that these particles are indistinguishable from electrons, which had been found by the English physicist Joseph John Thomson in 1897. Further exploration showed that the photoelectric impact addresses a cooperation among light and matter that can't be clarified by traditional physical science, which portrays light as an electromagnetic wave.

Applications of Photoelectric effect

Gadgets dependent on the photoelectric impact have a few advantageous properties, including creating a flow that is straightforwardly corresponding to light power and an extremely quick reaction time. One fundamental gadget is the photoelectric cell, or photodiode. Initially, this was a phototube, a vacuum tube containing a cathode made of a metal with a little work so electrons would be handily radiated. The current delivered by the plate would be accumulated by an anode held at an enormous positive voltage comparative with the cathode. Phototubes have been supplanted by semiconductor-based photodiodes that can recognize light, measure its power, control different gadgets as a component of enlightenment, and transform light into electrical energy. These gadgets work at low voltages, similar to their bandgaps, and they are utilized in modern cycle control, contamination observing, light discovery inside fiber optics broadcast communications organizations, sunlight based cells, imaging, and numerous different applications. Photoconductive cells are made of semiconductors with bandgaps that compare to the photon energies to be detected. For instance, visual openness meters and programmed switches for road lighting work in the noticeable range, so they are regularly made of cadmium sulfide. Infrared identifiers, for example, sensors for night-vision applications, might be made of lead sulfide or mercury cadmium telluride.

Photovoltaic gadgets commonly fuse a semiconductor p-n intersection. For sun oriented cell use, they are normally made of glasslike silicon and convert around 15% of the occurrence light energy into power. Sun based cells are regularly used to give somewhat limited quantities of force in exceptional conditions, for example, space satellites and remote phone establishments. Advancement of less expensive materials and higher efficiencies might make sunlight based power monetarily plausible for enormous scope applications. The photomultiplier tube is a profoundly touchy expansion of the phototube, first created during the 1930s, which contains a progression of metal plates called dynodes. Light striking the cathode discharges electrons. These are drawn to the first dynode, where they discharge extra electrons that strike the second dynode, etc. Later up to 10 dynode stages, the photocurrent is colossally enhanced that some photomultipliers can for all intents and purposes identify a solitary photon. These gadgets, or strong state renditions of similar affectability, are priceless in spectroscopy research, where it is generally expected important to quantify very powerless light sources. They are additionally utilized in shine counters, which contain a material that produces glimmers of light when struck by X-beams or gamma beams, coupled to a photomultiplier that counts the blazes and measures their force. These counters support applications, for example, recognizing specific isotopes for atomic tracer examination and distinguishing X-beams utilized in automated pivotal tomography (Feline) sweeps to depict a get segment through the body.

Photodiodes and photomultipliers likewise add to imaging innovation. Light speakers or picture intensifiers, TV camera cylinders, and picture stockpiling tubes utilize the way that the electron discharge from each point on a cathode is controlled by the quantity of photons showing up by then. An optical picture falling on one side of a cloudy cathode is changed over into a same "electron current" picture on the opposite side. Then, at that point, electric and attractive fields are utilized to concentrate the electrons onto a phosphor screen. Every electron striking the phosphor creates a blaze of light, causing the arrival of a lot additional electrons from the comparing point on a cathode straightforwardly inverse the phosphor. The subsequent heightened picture can be additionally improved by a similar cycle to create considerably more noteworthy enhancement and can be shown or put away. At higher photon energies the investigation of electrons radiated by X-beams gives data about electronic advances among energy states in iotas and atoms. It additionally adds to the investigation of specific atomic cycles, and it assumes a part in the compound examination of materials, since transmitted electrons convey a particular energy that is normal for the nuclear source. The Compton impact is additionally used to dissect the properties of materials, and in cosmology it is utilized to examine gamma beams that come from enormous sources.

Wave-Particle Duality

Is that molecule a wave? Or on the other hand is that wave a molecule? That is one of the inquiries that quantum physical science was made to tackle, since waves showed particle like properties while particles also display wave-like properties in the lab. Wave-Particle duality refers to the ability of some physical subtance (like light and electrons) of both wavelike and particle like attributes. Based on trial proof, German physicist Albert Einstein previously showed (1905) that light, which had been viewed as a type of electromagnetic waves, should likewise be considered as molecule like, limited in parcels of discrete energy. The perceptions of the Compton impact (1922) by American physicist Arthur Holly Compton could be clarified provided that light had a wave-molecule duality. French physicist Louis de Broglie proposed (1924) that electrons and other discrete pieces of issue, which up to that point had been imagined distinctly as material particles, additionally have wave properties like frequency and recurrence. Afterward (1927) the wave idea of electrons was tentatively settled by American physicists Clinton Davisson and Lester Germer and freely by English physicist George Paget Thomson. A comprehension of the correlative connection between the wave viewpoints and the molecule parts of a similar peculiarity was reported by Danish physicist Niels Bohr in 1928.

Heisenberg's uncertainty principle

One of the victories of quantum physical science is the Heisenberg Uncertainity Rule: Heisenberg guessed that you can't at the same time measure a molecule's position and energy precisely. This is one of the focal speculations that has annihilated old style physical science.

Harmonic Oscillator

Quantizing harmonic oscillators on the miniature level was one more victory of quantum physical science. Traditionally, harmonic oscillators can have any energy — yet not quantum precisely. It is an actual framework where some value wavers above and under a mean value with at least one trademark frequencies. Such frameworks frequently emerge when an opposite power results from

182

dislodging from a power nonpartisan position, and gets more grounded with respect to how much uprooting. For instance, pulling or pushing the finish of a spring from its rest position brings about a power pushing back toward the rest position. Releasing the spring from a place of strain brings about symphonious movement of the spring; the spring is presently a consonant oscillator. Different models incorporate a swinging pendulum, a vibrating violin string, or an electronic circuit that produces radio waves.

Schrödinger's Cat

Schrödinger's Cat is a psychological test that subtleties a few issues that emerge in the large scale world from considering the twist of electrons not really settled until you measure them. For instance, assuming you know the twist of one of a couple of recently made electrons, you know different must have the contrary twist. So assuming you separate two electrons by light years and afterward measure the twist of one electron, does the other electron's twist abruptly snap to the contrary esteem — even a good ways off that would take a sign from the principal electron years to cover? Interesting stuff! In quantum mechanics, Schrödinger's cat is a psychological study that represents a mystery of quantum superposition. In the psychological study, a theoretical cat might be viewed as all the while both alive and dead because of its destiny being connected to an irregular subatomic occasion that could conceivably happen. Schrödinger's Cat: a cat, a jar of toxic substance, and a radioactive source are set in a fixed box. If an inward screen (for example Geiger counter) distinguishes radioactivity (for example a solitary molecule rotting), the cup is broken, delivering the toxin, which kills the cat. The Copenhagen understanding of quantum mechanics suggests that sooner or later, the cat is at the same time alive and dead. However, when one examines the case, one sees the cat either alive or dead, not both alive and dead. This suggests the conversation starter of when precisely quantum superposition finishes and reality settle into one chance or the other. This psychological study was conceived by physicist Erwin Schrödinger in 1935, in a conversation with Albert Einstein,to show what Schrödinger considered the issues of the Copenhagen understanding of quantum mechanics. The situation is frequently highlighted in hypothetical conversations of the translations of quantum mechanics, especially in circumstances including the estimation problems.

Square Wells

Like harmonic oscillators, quantizing particles bound in square wells at the micro level was another triumph for quantum physics. Classically, particles in square wells can have any energy, but quantum physics says you can only have certain allowed energies. The finite square well (otherwise called the finite potential well) is an idea from quantum mechanics. It is an augmentation of the endless potential well, where a molecule is restricted to a "box", yet one which has limited potential "fence". In contrast to the limitless potential well, a likelihood related with the molecule is being found fresh. The quantum mechanical translation is not normal for the traditional understanding, where in the event that the all out energy of the molecule is not exactly the potential energy boundary of the walls, it can't be found fresh. In the quantum translation, there is a non-no likelihood of the molecule being fresh in any event, when the energy of the molecule is not exactly the potential energy obstruction of the fence.

Quantum Tunnelling

The quantum world is a wild one, where the apparently outlandish happens constantly: Insignificant objects isolated by miles are attached to each other, and particles can even be in two spots on the double. Yet, perhaps the most puzzling quantum superpower is the movement of particles through apparently invulnerable obstructions. How could particles go where, traditionally, they need more energy to go? For instance, how could an electron with energy 'E' go into an electrostatic field where you really want to have more than energy 'E'to infiltrate? The response was hypothesized with

Quantum Tunnelling. Quantum tunnelling refers to a situation where an atom or a subatomic molecule can show up on the other side of an obstruction that ought to be impossible for the molecule or particle to pass through. Maybe you were strolling and experienced a 10-foot-tall (3 meters) fence stretching out as may be obvious. Without a stepping stool climbing abilities, the fence would make proceeding to the other side a task that you can't accomplish. Nonetheless, in the quantum world, it is uncommon, yet conceivable, for a molecule or electron to just "show up" on the opposite side, as though a passage had been burrowed through the fence. "Quantum tunnelling is to be sure one of the most astounding of quantum phenomena. An example is the tunnelling of an old style wave-molecule affiliation, transient wave coupling (the use of Maxwell's wave-condition to light) and the utilization of the non-dispersive wave-condition from acoustics applied to "waves on strings". Quantum tunnelling assumes a fundamental part in actual peculiarities, for example, atomic fusion.It has applications in the tunnel diode,quantum processing, and in the filtering tunnelling magnifying lens. Quantum tunnelling is projected to make actual cutoff points to the size of the semiconductors utilized in microelectronics, because of electrons having the option to burrow past semiconductors that are excessively little.

Postulating spin

The Stern Gerlach analyze results couldn't be clarified without hypothesizing turn, one more victory of quantum physical science. This test sent electrons through an attractive field, and the traditional expectation is that the electron stream would make one spot of electrons on a screen — however there were two (corresponding to the two up and down spins). In 1921, Otto Stern and Walter Gerlach carried out a test which showed the quantization of electron spin into two directions. This made a significant commitment to the improvement of the quantum theory of atom. The real test was done with a light emission molecules from a hot broiler since they could be promptly distinguished utilizing a visual emulsion. The silver particles permitted Stern and Gerlach to concentrate on the attractive properties of a solitary electron on the grounds that these iotas have a solitary external electron which moves in the Coulomb potential brought about by the 47 protons of the core protected by the 46 inward electrons. Since this electron has zero orbital rakish force (orbital quantum number l=0), one would expect there to be no communication with an outer attractive field. Stern and Gerlach coordinated the light emission particles into an area of non uniform attractive field . An attractive dipole second will encounter a power corresponding to the field slope since the two "posts" will be dependent upon various fields.

Traditionally one would expect all potential directions of the dipoles so a consistent smear would be created on the visual plate, yet they observed that the field isolated the pillar into two unmistakable parts, showing only two potential directions of the attractive snapshot of the electron.

Be that as it may, how does the electron get a magnetic moment assuming it has zero angular momentum and in this way delivers no "current circle" to create a magnetic moment? In 1925, Samuel A. Goudsmit and George E. Uhlenbeck proposed that the electron had an inborn precise energy, free of its orbital qualities. In old style terms, a bundle of charge could have an attractive second assuming it were turning with

193

the end goal that the charge at the edges delivered a compelling current circle. This sort of thinking prompted the utilization of "electron spin" to depict the characteristic angular momentum.

Contrasts between Newton's Laws and Quantum physical science

In traditional (classical) physical science, bound particles can have any energy or speed, yet that is false in quantum physical science. What's more in old style physical science, you can decide both the position and momentum of particles precisely, which isn't accurate in quantum physical science (because of the Heisenberg Uncertainty principle). Furthermore, in quantum physical science, you can superimpose states on one another, and have particles burrow into regions that would be traditionally outlandish.

Expanding on the work done by Galileo and others, Newton Stated his laws of motion in 1686. According to Newton:

• A body remains in a state of rest or uniform motion in a straight line unless it is acted upon by an external force.

• The rate of change of momentum of a body is directly proportional to the external applied force and always take the direction of the applied force.

• For every action, an equal and opposite reaction exist.

Classical (Newtonian)mechanics is totally deterministic: Given the specific positions and speeds of all particles at a given time, alongside the capacity, one can work out the future (and past) positions and speeds of all particles at some other time. The advancement of the framework's positions and momenta through time is regularly alluded to as trajectory. Various exploratory perceptions in the last part of the 1800's and mid 1900's constrained physicists to look past Newton's laws of movement for a broader hypothesis. It couldn't be any more obvious, for instance, the conversation of the hotness limit of solids. It had become progressively certain that electromagnetic radiation had molecule like properties

196

notwithstanding its wave-like properties like diffraction and obstruction. Planck displayed in 1900 that electromagnetic radiation was discharged and consumed from a dark body in discrete quanta, each having energy corresponding to the recurrence of radiaion. In 1904, Einstein conjured these quanta to clarify the photograph electric impact. So in specific situations, one should decipher electromagnetic waves as being comprised of particles. In 1924 de Broglie stated that matter additionally had this double nature: Particles can be wavey. To make a long and astonishing story short, this prompted the definition of Shrödinger's wave condition for matter. this fractional differential equation is hard to manage and by and large difficult to settle analytically.

Quantum mechanics is in this manner not deterministic, however probabilistic. It constrains us to leave the thought of unequivocally characterized directions of particles through reality. All things considered; we should talk as far as probabilities for elective framework arrangements. To explain these ideas, consider two significant triumphs for the quantum hypothesis, forecasts of the discrete energy levels of the harmonic oscillator and the hydrogen atom.

Discrete Spectra of Molecules

Demonstrating the quantized idea of particles and orbitals is one more victory of quantum material science. Incidentally, electrons can't have just any energy in a particle, however are just permitted specific quantized energy levels — and that was one of the establishments of quantum physical science. Spectroscopy is a part of science that concerns the examination and estimation of spectra created when matter communicates with or emits electromagnetic radiation. In straightforward terms, spectroscopy is the scattering of light into its constituent colors.

We realize that when a matter is presented to electromagnetic radiation, the electromagnetic range of a progression of frequencies produces. Molecules retain specific frequencies to the higher electronic, vibrational, and rotational energy levels. So, the series of frequencies, a particle ingests emits a particular molecular spectrum. A particular molecular spectrum lies in a particular district of the electromagnetic spectrum.

Molecular spectra are of three types. They include the following:

1.   Pure Rotational spectra: At the point when a particle retains a lower measure of energy that it makes a change starting with one rotational level then onto the next inside the equivalent vibrational level, Rotational spectra are detectable in the phantom district of Far Infrared and Microwaves. Also, the energies in these ghastly locales are minuscule. Along these

lines, we call rotational spectra the microwave spectra.

2.   Vibrational Rotational Spectra: At the point when an atom retains adequate energy that influences the change of a particle starting with one vibrational level then onto the next inside the equivalent electronic level. Therefore, for this situation, both rotational and vibrational progress happens. This is the way we acquire vibrational rotational spectra. The vibrational spectra are discernible in the Close Infrared spectral area. We call the vibrational rotational spectra the Infrared spectra.

3.   Electronic Band spectra: At the point when the intriguing energy of the radiation is adequately enormous to cause the progress of a

particle starting with one electronic level then onto the next electronic level. This progress is joined by both rotational and vibrational level changes. Additionally, for each vibrational change, a bunch of firmly divided lines appear. Because of the relating rotational level changes, these firmly separated lines are known as bands. Henceforth, we call it the electronic band spectra.

Conclusion

This Book was written with the sole aim of aiding quick and easy learning of the fundamentals of quantum Physics especially for Beginners. It elaborate on the concept of quantum Physics depicting it in clear unambiguous term for the benefit of users. Quantum Physics is the investigation of the conduct of matter and energy at the sub-atomic, nuclear, atomic, and surprisingly more modest tiny levels. In the mid twentieth century, researchers found that the laws administering plainly visible items don't work something similar in such little domains.

Quantum" comes from the Latin signifying "the amount." It alludes to the discrete units of matter and energy that are anticipated by and seen in quantum material science. Indeed, even reality, which give off an impression of being very persistent, have the littlest potential qualities. As researchers acquired the innovation to quantify with more prominent accuracy, bizarre peculiarities was noticed. The introduction of quantum material science is credited to Max Planck's 1900 paper on blackbody radiation. Improvement of the field was finished by Max Planck, Albert Einstein, Niels Bohr, Richard Feynman, Werner Heisenberg, Erwin Schroedinger, and other illuminating presence figures in the field. Amusingly, Albert Einstein disapproved of quantum mechanics and went after it for a long time to invalidate or change it.

In the domain of quantum physical science, noticing something really impacts the actual cycles occurring. Light waves carry on like particles and particles behave like waves (called wave particle duality). Matter can move between various spots without traveling through the interceding space (called quantum tunnelling).

Quantum Physics for beginners

Data moves quickly across tremendous distances. Indeed, in quantum mechanics we find that the whole universe is really a progression of probabilities. Luckily, it separates when managing huge items, as shown by the Schrodinger's cat psychological test. One of the key ideas is quantum entrapment, which depicts a circumstance where different particles are related so that estimating the quantum condition of one molecule additionally puts imperatives on the estimations of different particles. This is best exemplified by the EPR (Einstein podoslky Rosen) Pradox. However initially a psychological test, this has now been affirmed tentatively through trial of something known as Bell's Hypothesis.

Quantum optics is a part of quantum material science that shines essentially on the conduct of light, or photons. At the degree of quantum optics, the conduct of individual photons has a course on the outcoming light, instead of traditional optics, which was created by Sir Isaac Newton. Lasers are one application that has emerged from the investigation of quantum optics. Quantum electrodynamics (QED) is the investigation of how electrons and photons associate. It was created in the last part of the 1940s by Richard Feynman, Julian Schwinger, Sinitro Tomonage, and others. The forecasts of QED with respect to the dissipating of photons and electrons are exact to eleven decimal places. Unified field theory is an assortment of exploration ways that are attempting to accommodate quantum physical science with Einstein's hypothesis of general relativity, frequently by attempting to combine the key powers of physical science. A few sorts of unified theory include the following:

- Quantum Gravity

- Circle Quantum Gravity

- String Hypothesis/Superstring Hypothesis/M-Hypothesis

- Great unified Hypothesis

- Supersymmetry

Quantum phyics is sometimes referred to as quantum mechanics or quantum field hypothesis. It additionally has different subfields, as examined above, which are here and there utilized conversely with quantum physical science, however quantum physical science is really the more extensive term for these disciplines. Significant Discoveries, Investigations, and Essential Clarifications include:

- Black Body Radiation
- Photoelectric Impact
- Wave-Particle Duality
- De Broglie Theory
- The Compton effect
- Heisenberg Uncertainty principle
- The Copenhagen Understanding
- Schrodinger's cat

- EPR (Eistein Podolsky Rosen)paradox

And many more to mention but few.

References

1.   R. Arnowitt, S. Deser, C.W. Misner, "The Dynamics of General Relativity" in Gravitation: An Introduction to Current Research, ed. by L. Witten (Wiley, New York, 1962), [grqc/0405109]

2.   J.N. Goldberg, J. Lewandowski, C. Stornaiolo: "Degeneracy in Loop Variables", Comm. Math. Phys. 148, (1992),377;

3.   T. Jacobson, J.D. Romano: "The Spin Holonomy Group in General Relativity", Commun. Math. Phys. 155, (1993), 261

4.   T. Thiemann, "The Phoenix Project: Master Constraint Programme for Loop Quantum Gravity", [gr-qc/0305080]

5.   C. Rovelli, "Loop Quantum Gravity", Living Rev. Rel. 1 (1998) 1, [gr-qc/9710008]

6.   T. Thiemann,"Lectures on Loop Quantum Gravity", in Quantum Gravity: From Theory to Experimental Search Proceedings, Bad Honnef, Germany 2002, eds D. Giulini, C. Kiefer,

7.   C. L¨ammerzahl, LNP 631 (Springer, Berlin 2003), [gr-qc/0210094]

A.   Ashtekar,   J.   Lewandowski, "Background Independent Quantum Gravity: A Status Report", Class. Quant. Grav. 21 (2004) R53; [gr-qc/0404018]

8.   L. Smolin, "An Invitation to Loop Quantum Gravity", [hep-th/0408048]

9.   C. Rovelli, "Quantum Gravity" (Cambridge University Press, Cambridge 2004)

10.   T. Thiemann, "Modern Canonical Quantum General Relativity", (Cambridge University Press, Cambridge 2005), [gr-qc/0110034]

11. M. Henneaux, C. Teitelboim, "Quantization of Gauge Systems" (Princeton University Press 1992)

12. C. Lanczos, "The variational principles of mechanics" (University of Toronto Press 1970)

A. Wipf, "Hamilton's Formalism for Systems with Constraints", in Canonical Gravity: From Classical to Quantum, Proceedings, Bad Honnef, Germany 1993, ed. by J. Ehlers and H. Friedrich, LNP 434 (Springer, Berlin 1994)

13. D.M. Gitman, I.V. Tyutin, "Quantization of Fields with Constraints", (Springer, Berlin 1990)

14. C. Rovelli, "Quantum mechanics without time: A model", Phys. Rev. D42, (1990), 2638

15. C. Rovelli, "Time in quantum gravity: An hypothesis", Phys. Rev. D43, (1991), 442

16. C. Rovelli, "Is There Incompatibility Between the Ways Time is Treated in General Relativity and in Standard Quantum Mechanics" in Conceptional Problems in Quantum Gravity, ed. by A. Ashtekar and J. Stachel (Birkh¨auser, Boston, 1991),

17. C. Rovelli, "What is observable in classical and quantum gravity?", Class. Quant. Grav. 8,

18. (1991) , 1895

19.   C. Rovelli, "Partial Observables", Phys. Rev. D65, (2002), 124013, [gr-qc/0110035] ; C. Rovelli, "Quantum Gravity", (Cambridge University Press, Cambridge 2004)

20.   P.G. Bergmann, "Introduction of 'True Observables' into the Quantum Field Equations", Nuovo Cimento 3 (1956)

21.   P.A.M. Dirac, "Lectures in Quantum Mechanics", (Yeshive Univ., New York, 1964)

A.   Ashtekar, R.S. Tate, "An algebraic extension of Dirac quantization: Examples", J. Math. Phys. 35, (1994), 6434, [gr-qc/9405073]

A.   Higuchi, "Quantum linearization instabilities of de Sitter spacetime II", Class. Quant. Grav. 8, (1991), 1983

A.   Higuchi, "Linearized quantum gravity in flat space with toroidal topology", Class. Quant. Grav. 8, (1991), 2023

22.   N.P. Landsman, "Rieffel induction as generalized quantum Marsden–Weinstein reduction", J. Geom. Phys. 15, (1995), [hep-th/9305088]

23.   D. Giulini, D. Marolf , "A Uniqueness Theorem for Constraint Quantization" Class. Quant. Grav. 16, (1999), 2489, [gr-qc/9902045];

24.   D. Giulini, D. Marolf, "On the Generality of Refined Algebraic Quantization", Class. Quant. Grav. 16, (1999), [gr-qc/9812024]

A.   Gomberoff, D. Marolf, "On Group Averaging for SO(n,1)",Int. J. Mod. Phys. D8, (1999), 519 , [gr-qc/9902069]

25.   D. Marolf, "The Spectral Inner Product for Quantum Gravity", in Proceedings of the VII– th Marcel–Grossman Conference, eds. R. Ruffini, M. Keiser, (World Scientific, Singapore 1995), [gr-qc/9409036]

26.   J. Klauder, "Universal Procedure for Enforcing Quantum Constraints", Nucl. Phys. B547, (1999), [hep-th/9901010];

27.   J.   Klauder,   "Quantization   of Constrained Systems", Lect. Notes Phys. 572 (2001) 143-182,

28.   [hep-th/0003297]

A.   Kempf,   J.   Klauder,   "On   the Implementation   of   Constraints   through Projection Operators", J. Phys. A34 (2001), 1019, [quant-ph/0009072]

29.   K.V.   Kuchaˇr,   "A   Bubble-Time Canonical Formalism for Geometrodynamics", J. Math. Phys. 13, (1972), 768

30.   K.V.   Kuchaˇr,   "Time   and Interpretations of Quantum Gravity", in Proceedings of the 4th Canadian Conference on General Relativity and Relativistic Astrophysics, eds. G. Kunstatter, D. Vincent and J. Williams (World Scientific, Singapore, 1992)

31.   K.V.   Kuchaˇr,   "Canonical Quantization of Cylindrical Gravitational Waves", Phys. Rev. D4, (1971), 955

32.   J.D. Brown, K.V. Kuchaˇr, "Dust as a Standard of Space and Time in Canonical Quantum Gravity", Phys. Rev. D51 (1995), 5600, [gr-qc/9409001]

33.   K.V. Kuchaˇr, C.G. Torre, "Gaussian Reference Fluid and Interpretation of Quantum Geometrodynamics", Phys. Rev. D43, (1991), 419

34.   C.J. Isham, "Canonical Quantum Gravity and the Question of Time", in Canonical Gravity: From Classical to Quantum, Proceedings, Bad Honnef, Germany 1993, ed. by J. Ehlers and H. Friedrich, LNP 434 (Springer, Berlin 1994)

35.   C.G. Torre: "Is general relativity an 'already parametrized' theory?", Phys. Rev. D 46, (1992), 3231

36.   G. Roepstorff, "Path Integral Approach to Quantum Physics: An Introduction", (Springer, Berlin 1994)

37. K.V. Kucha˘r, "Canonical Quantization of Generally Covariant Systems", in Highlights in Gravitation and Cosmology, Proceedings of the International Conference on Gravitation and

38. Cosmology, Goa, India, 1987 eds. B.R. Iyer et al., (Cambridge University Press, Cambridge 1988)

39. C.G. Torre, "A complete set of observables for cylindrically symmetric gravita tional fields", Class. Quantum Grav. 8, (1991), 1895

40. P.G. Bergmann: "General Theory of Relativity" in Encyclopedia of Physics, Vol. IV: Principles of Electrodynamics and Relativity, ed. by S. Fl¨ugge (Springer, Berlin 1962)

41. B.S. DeWitt, "The quantization of geometry", in Gravitation: An Introduction to Current

42. Research, ed. by L. Witten (Wiley, New York 1962)

43. K.V. Kuchaˇr, "Geometry of Hyperspace", J. Math. Phys. 17, (1977), 777

44. K.V. Kuchaˇr, "Kinematics of Tensor Fields in Hyperspace", J. Math. Phys. 17, (1977) 792 [34] V. I. Arnold (Ed.), "Dynamical Systems III" (Springer, Berlin 1988)

45. M. Reed, B. Simon: "Methods of Modern Mathematical Physics I: Functional Analysis" (Academic Press, London 1980)

46. S.L. Lykhaovich, R. Marnelius, "Extended Observables in Hamiltonian

Theories with COnstraints", Int. J. Mod. Phys. A16, (2001), 4271, [hep-th/0105099]

47.   L. Smolin, "Time, Measurement and Information Loss in Quantum Cosmology" in College Park 1993, Directions in general relativity, Vol. 2, ed. by B.L. Hu et. al. (Cambridge Univ. Press, 1993), [gr-qc/9301016]

48.   T. Thiemann, "Reduced Phase Space Quantization and Dirac Observables", [gr-qc/0411031]

49.   R. M. Wald, "General Relativity", (Chicago Univ. Press, 1984)

50.   C.G.   Torre,   "Gravitational Observables and Local Symmetries", Phys. Rev. D48, (1993), 2373, [gr-qc/9306030]

A.   Ashtekar, C.J. Isham, "Representations of the Holonomy Algebras of Gravity and NonAbelean Gauge Theories", Class. Quantum Grav. 9 (1992) 1433, [hep-th/9202053]

A.   Ashtekar,   J.   Lewandowski, "Representation theory of analytic Holonomy C4 algebras", in Knots and Quantum Gravity eds. J. Baez, (Oxford University Press, Oxford 1994)

51.   H. Sahlmann, "When do Measures on the Space of Connections Support the Triad Operators of Loop Quantum Gravity?", [gr-qc/0207112];

52.   H. Sahlmann, "Some Comments on the Representation   Theory   of   the   Algebra Underlying

223

53. Loop Quantum Gravity", [gr-qc/0207111];

54. H. Sahlmann, T. Thiemann, "On the Superselection Theory of the Weyl Algebra for Diffeomorphism Invariant Quantum Gauge Theories", [gr-qc/0302090];

55. H. Sahlmann, T. Thiemann, "Irreducibility of the Ashtekar-Isham-Lewandowski Representation", [gr-qc/0303074];

A. Okolow, J. Lewandowski, "Diffeomorphism Covariant Representations of the Holonomy Flux Algebra", [gr-qc/0302059];

B. Fleichhack, "Representations of the Weyl Algebra in Quantum Geometry", [mathph/0407006];

56.   J.   Lewandowski,   A.   Okolow,   H. Sahlmann, T. Thiemann, "Uniqueness of Diffeomorphism Invariant Representations of the   Holonomy   –   Flux   Algebra",   [gr-qc/0504147]

57.   T.   Thiemann,   "Anomaly-free Formulation   of   non-perturbative,   four-dimensional Lorentzian Quantum Gravity", Physics   Letters   B380   (1996)   257-264,   [gr-qc/9606088];

58.   T.   Thiemann,   "Quantum   Spin Dynamics (QSD)", Class. Quantum Grav. 15 (1998) 83973, [gr-qc/9606089]; "II. The Kernel of the Wheeler-DeWitt Constraint Operator", Class. Quantum Grav. 15 (1998) 875-905, [gr-qc/9606090];   "III.   Quantum   Constraint

Algebra and Physical Scalar Product in Quantum General Relativity", Class. Quantum Grav. 15 (1998) 1207-1247, [gr-qc/9705017]; "IV. 2+1 Euclidean Quantum Gravity as a model to test 3+1 Lorentzian Quantum Gravity", Class. Quantum Grav. 15 (1998) 1249-1280, [grqc/9705018]; "V. Quantum Gravity as the Natural Regulator of the Hamiltonian Constraint of Matter Quantum Field Theories", Class. Quantum Grav. 15 (1998) 1281-1314, [gr-qc/9705019]; "VI. Quantum Poincar´e Algebra and a Quantum Positivity of Energy

59.	Theorem for Canonical Quantum Gravity", Class. Quantum Grav. 15 (1998) 1463-1485,

60.   [gr-qc/9705020]

61.   L. Smolin, "The Classical Limit and the Form of the Hamiltonian Constraint in Non–Perturbative Quantum General Relativity", [gr-qc/9609034]

62.   J. Lewandowski, D. Marolf, "Loop Constraints: A Habitat and their Algebra", Int. J. Mod. Phys. D7, (1998), 299, [gr-qc/9710016]

63.   R. Gambini, J. Lewandowski, D. Marolf, J. Pullin, "On the Consistency of the Constraint Algebra in Spin Network Gravity", Int. J. Mod. Phys. D7, (1998), 97, [gr-qc/9710018]

64.   T. Thiemann, "Quantum Spin Dynamics (QSD): VII. Symplectic Structures and Continuum Lattice Formulations of Gauge

Field Theories", Class. Quant. Grav. 18, (2001), 3293 [hep-th/0005232];

65.   T. Thiemann, "Gauge Field Theory Coherent States (GCS): I. General Properties", Class.

66.   Quant. Grav. 18, (2001), 2025 [hep-th/0005233];

67.   T. Thiemann, "Complexifier Coherent States for Canonical Quantum General Relativity", [gr-qc/0206037]

68.   T. Thiemann, O. Winkler, "Gauge Field Theory Coherent States (GCS): II. Peakedness Properties", Class. Quant. Grav. 18, (2001), 2561, [hep-th/0005237]; "III. Ehrenfest Theorems", Class. Quantum Grav. 18, (2001), 4629, [hep-th/0005234]; "IV. Infinite Tensor

Product and Thermodynamic Limit", Class. Quantum Grav. 18, (2001), 4997, [hep-th/0005235]; H. Sahlmann, T. Thiemann, O. Winkler, "Coherent States for Canonical Quantum General Relativity and the Infinite Tensor Product Extension", Nucl. Phys. B606, (2001) 401, [gr-qc/0102038]

69.   B. Dittrich, T. Thiemann, "Testing the Master Constraint Programme for Loop Quantum Gravity I. General Framework", [gr-qc/0411138]

70.   M.S. Birman, M. Z. Solomjak, "Spectral Theory of Self – Adjoint Operators in Hilbert

71.   Space", (D. Reidel Publishing Company, Doordrecht, 1987)

72. A. Galindo, P. Pascual, "Quantum Mechanics I", (Springer Verlag, Berlin 1990)

73. M. Reed, B. Simon, "Methods of Modern Mathematical Physics II: Fourier Analysis, SelfAdjointness", (Academic Press, New York 1975)

74. G. Teschl, "Schr¨odinger Operators", lecture notes for a cours at the University Vienna 1999 and 2002, available at http://www.mat.univie.ac.at/~gerald/ftp/boo k~schroe/

75. R. Howe,E. C. Tan, "Non-Abelian Harmonic Analysis – Applications of SL(2,R)", (Springer, New-York 1992)

76.   A. Paterson, "Amenability", (American Mathematical Society, Providence, Rhode Island,

77.   1988)

78.   J.P. Pier, "Amenable Locally Compact Groups", (Wiley, New York, 1984)

79.   A. Carey, H. Grundling, "Amenability of the Gauge Group", [math-ph/0401031]

80.   R. Howe, "Dual pairs in physics: Harmonic oscillators, photons, electrons, and singletons", in *Lectures in Appl. Math., Vol. 21*, (Amer. Math. Soc., Providence, RI, 1979)

81.   M. Montesinos, C. Rovelli, T. Thiemann, "$SL(2,R)$ model with two Hamiltonian constraints", Phys. Rev. **D60**, (1999), 044009, [gr-qc/9901073]

82.   J. Louko, C. Rovelli, "Refined algebraic quantization in the oscillator representation of $SL(2,R)$", J. Math. Phys. **41**, (2000), 132, [gr-qc/9907004]

83.   J. Louko, A. Molgado, "Group averaging in the (p,q) oscillator representation of $SL(2,R)$", J. Math. Phys. **45**, (2004), 1919 , [gr-qc/0312014]

84.   M. Trunk, "An $SL(2,R)$ model of constrained systems: Algebraic constrained quantization", University of Freiburg preprint THEP 99/3, [hep-th/9907056]

85.   R. Gambini, R.A. Porto, "Relational time in generally covariant quantum systems: Four models", Phys. Rev. **D63**, (2001), 105014 , [gr-qc/0101057]

86. H. Neunh¨offer, "Uber Kronecker-Produkte irreduzibler Darstellungen von¨ $SL(2,R)$", Sitzungsberichte der Heidelberger Akademie der Wissenschaften, Mathematischnaturwissenschaftliche Klasse, Jahrgang 1978, 167

87. J. Repka, "Tensor products of unitary representations of $SL(2,R)$", Amer. J. Math. **100(4)**, (1978), 747

88. J.E. Marsden, P.R. Chernoff, "Properties of Infinite Dimensional Hamiltonian Systems", Lecture Notes in Mathematics, (Springer-Verlag, Berlin, 1974)

89. T. Thiemann, "Generalized Boundary Conditions for General Relativity for the Asymptotically Flat Case in Terms of Ashtekar's

Variables", Class. Quant. Grav. **12**, (1995), 181, [gr-qc/9910008]

90.   A. Ashtekar, C. Rovelli, L. Smolin, "Gravitons and Loops", Phys. Rev. **44**, (1991), 1740, [hep-th/9202054]

91.   A. Ashtekar, J. Lee, "Weak Field Limit of General Relativity: A New Hamiltonian Formulation", Int. J. Mod. Phys. **D3**, (1994), 675

92.   O.Y. Shvedov, "BRST-BFV, Dirac and Projection Operator Quantizations: Correspondence of States", [hep-th/0106250];

93.   O.Y. Shvedov, "Refined Algebraic Quantization of Constrained Systems with Structure Functions", [hep-th/0107064];

94.   O.Y. Shvedov, "On Correspondence of BRST-BFV, Dirac and Refined Algebraic Quantizations of Constrained Systems", Annals Phys. **302**, (2002), 2, [hep-th/0111270]

95.   P. Hajiˇcek, K.V. Kuchaˇr, "Constraint Quantization of parametrized relativistic gauge systems in curved spacetimes", Phys. Rev. **D41**, (1990), 1091;

96.   P.   Hajiˇcek,   K.V.   Kuchaˇr, "Transversal affine connection and quantization of constrained systems", Journ. Math. Phys. **31** (1990) 1723

97.   V. Bargmann, "Irreducible unitary representations of the Lorentz group", Ann. of Math. **48**,(1947), 568

98.   B.G. Adams, J. Cizek, J. Paldus, "Lie Algebraic Methods and Their Applications to Simple Quantum Systems", Adv. Quantum Chem. **19**,(1988),1

99.   R. Howe, "On some results of Strichartz and of Rallis and Schiffman", J. Funct. Anal. **32**, (1979), 297

100.   P. Haji˘cek, "Comment on 'Time in quantum gravity: An hypothesis' ", Phys. Rev. **D44**, (1991), 1337

101.   C. Rovelli, "Quantum evolving constants. Reply to 'Comment on 'Time in quantum gravity:

102.   An hypothesis' ' " , Phys. Rev. **D44**, (1991), 1339